Why Claims That Humans Are Not Causing Global Warming and Climate Change Are Wrong

by

David Leithauser

Copyright © 2018 David Leithauser

All Rights Reserved

Table of Contents

Introduction 3
How Global Warming Works 7
The World is Not Warming 13
The Temperature is Not Rising Steadily 18
It Is All a Hoax 21
Climate Scientists Changed "Global Warming" to "Climate Change" 27
There Is No Consensus 29
Scientists Said the World Was Cooling in the 1970's 35
Scientists Cannot Even Predict the Weather 38
The Models Are Not Accurate 41
The Climate Has Always Changed 44
We Are Coming Out of an Ice Age 46
It's The Sun 49
CO2 Lags Temperature 53
It's Cold in (fill in the blank) 56
Earth Has Had Lower Temperatures with Higher CO2 58
Antarctica Is Gaining Ice 60
Sea Levels Are Not Rising 62
Melting Ice in Water Does Not Raise Water Levels 65
CO2 is a Trace Gas 67
It Is the Water, Not the CO2 69
Humans are Not the Main Contributors of CO2 70
CO2 is Plant Food 73
Global Warming Will Be Good 77
A Few Degrees Will Not Make a Difference 82
The World Survived Global Warming Before 85
Stopping Global Warming Would Destroy the Economy . 87
There is Probably Nothing You Can Do 98

Introduction

Global warming should not be controversial. It is not some new, untested theory. The science is well established. The scientist Eunice Foote first reported her discovery that CO2 (carbon-dioxide) traps heat from the sun, keeping the Earth warmer, in 1856. In 1896, the scientist Arrhenius first specifically warned that increasing the level of CO2 in the atmosphere could increase global temperatures. Since that time, scientists have continued to run more and more tests and experiments designed to increase our understanding of exactly what gases increase global warming, how much each gas does, and every other aspect of how global warming works. All of these studies have confirmed that CO2 and some other gases (such as methane) are increasing the global temperature as humans have increased the concentration of these gases in the atmosphere through burning fossil fuels and some other activities.

So why do some people claim that the science is not settled, or even totally deny it? The solution to preventing global warming is to stop, or at least strongly reduce, our emissions of CO2 and other greenhouse gases, which means significantly reducing our use of fossil fuels like coal, oil, and natural gas. This would significantly cut into the profits of the very rich and powerful people who sell these fossil fuels, and they are not happy with that idea. They have therefore launched massive propaganda campaigns to convince the public that global warming and climate change are not happening at all, or that it is not caused by humans, or any other argument they can come up with to convince people that we should not switch to clean, renewable energy sources like solar or wind power. They are using the same deceptive tactics that the tobacco companies used to try to convince people that smoking was not unhealthy and even that cigarettes are not addictive, the lead industry used to try to convince people that lead fumes from leaded gasoline are not toxic, and a variety of other companies have used to delay any effort to stop the sale of their dangerous products. They have created fake think tanks and "research" organizations whose purpose is not to actually discover the truth about global warming, but to come up with deceptive but convincing sounding arguments to convince people that man-made (technically referred to as anthropogenic) global warming and climate change are not happening. These claims use a variety of tactics. These include outright lies with no basis in truth, cherry picking data, stating true facts while omitting the reasons why these facts do not disprove global warming, deliberately misquoting people who talk about global warming to make them seem ridiculous, creating bizarre and scary sounding conspiracy theories to scare people into not even listening to scientists who are trying to warn people about the dangers of global warming, and various other tactics.

This does not mean that everyone who has doubts about man-made global warming is engaged in deliberate deception. The science that explains how man-made global warming works can be highly technical, and hard for the layman to understand in detail. It is easy to be deceived by a slick advertising campaign by the fossil fuel companies that uses simple and reasonable sounding pseudoscience to make it seem that the science is questionable. That is why I am writing this book, to simplify the science and explain why these deceptions are false.

To illustrate a simple example of the tactic of stating true facts while omitting the reasons why these facts do not disprove something, I can use an example of a claim by the tobacco companies. A few years before all the publicity that smoking causes cancer, there was a significant increase in the number of women who took up smoking, largely due to an effort by the cigarette companies to make it seem cool for women to smoke.

When the reports that smoking caused cancer came out, the cigarette companies asked why there had been no reports of an increase in cancer among women. They said that there should be a lot more women getting cancer now that they were smoking if smoking really causes cancer. The cigarette companies knew full well that it takes about five years of smoking to induce cancer, that it would take several more years for the cancer to show up, and several more years for statistics to be gathered to show the increase. There had not been time for these things to all happen when the reports of smoking causing cancer came out. However, the tobacco companies used the true fact that there had not yet been a reported increase in cancer among women to create the false impression that smoking must not be causing cancer, thanks to a lack of detailed understanding of the science among the general public. It all sounds so reasonable. In the same way, the fossil fuel companies and their allies (mostly politicians who receive large campaign contributions from the fossil fuel companies) can make arguments that sound quite plausible to the general public, even though those arguments have been totally debunked by scientists who have spent their lives studying this issue.

Note that this book is not intended to be a rigorous or detailed technical explanation of how global warming and climate change work. This book is intended to familiarize the general public with why common claims that there are doubts about whether global warming is happening and is caused by humans are false. I will try to explain clearly and simply why the deceptive claims that humans are not changing the climate are not true. I will not go into a lot of technical details that would most likely put most people to sleep, or at least cause your attention to wander. I will certainly avoid a lot of math, since many people hate math. I will just attempt to explain the basic fallacies and deception in each of the false claims put out there designed to deliberately confuse people, and allow you to make up your own mind along the way.

I will start out in the first chapter explaining the basics of how global warming and climate change work. In each of the following chapters, I will describe one misconception or deceptive claim about why man-made global warming or climate change are not happening and explain why it is not valid. Each chapter will be titled with the claim that it disproves or explains.

In some places, I will insert Web addresses of some online resources and articles that elaborate on some of the statements made in this book. However, since the Web is dynamic and Web pages or even entire sites can be taken down, I cannot guarantee how long after this book is published any of these sites will remain active.

I will also clarify here that I occasionally go back and update some of the data in this book and rerelease it without changing the copyright notice or some other references. Therefore, do not be confused if you see data in this book later than the copyright notice of 2018.

One last note. One claim often made is that people who try to warn people of the dangers of global warming and climate change (I will, by the way, explain the difference between these in this book) are trying to profit from the issue (one of the conspiracy theories). In order to reassure people that this is not my motive, I will donate 100% of the profits from this book to charities that combat global warming. In fact, I will not only donate 100% of the profits from this book, I will match the first $1,000 per year each year with my own money, so I will actually be losing money on this book.

Chapter 1

How Global Warming Works

In this chapter, I will describe the basic science of global warming. The details of how global warming works can get very detailed and involve a lot of math and detailed physics. I will try to keep such details to a minimum in this chapter, and just give you a layman's version. I will have to go over a little basic physics, but I will try to simplify it a little and not get too technical for the benefit of readers who are not climatologists or physicists.

We all know that the Earth gets its warmth from the sun. The sun emits a wide range of frequencies of light. Some of these frequencies are visible, and are what we normally call light. Some are too high frequency for us to see. Ultraviolet light is too high to see, but does cause sunburn. Infrared light is too low frequency to see, but we feel it as warmth. In this book, I will refer to all of these frequencies as light for simplicity (light is so much simpler than the technical term, electromagnetic radiation), even the frequencies that we cannot see such as infrared. It is important to understand also that even within the range of light such as infrared, there is a range of frequencies, as I will discuss shortly.

Sunlight strikes the ground and water, and is absorbed, causing these to warm up. The ground and water then warm the surrounding air by contact. Some frequencies of light are absorbed more by some things, and some frequencies are absorbed more by others. A black object absorbs a wider range of frequencies than a light-colored object, and therefore absorbs more energy and gets hotter when exposed to sunlight.

Most of the gases in our atmosphere are transparent to most frequencies of light, including infrared light. The light passes right through the air without much of the light being absorbed, and therefore the gases do not heat up much directly from the sunlight. Oxygen and nitrogen, which make up just over 99% of the air, are especially transparent. However, some gases do absorb or reflect some frequencies of light. Common examples of gases that absorb some frequencies of infrared light are carbon-dioxide (CO_2), methane (CH_4), ozone (O_3), nitrous oxide (N_2O), and CFCs (chlorofluorocarbons, synthetic gases used in air conditioners and refrigerators). These gases tend to especially absorb the very low-frequency infrared light. These heat-trapping gases are called greenhouse gases.

So far, I have been talking about objects absorbing light energy and getting hotter. However, when any object gets warm, it actually starts radiating off heat too. You have probably seen pictures of red-hot metal glowing. Actually, the old-fashioned incandescent light bulbs work by running an electric current through a wire, making it so hot that it glows brightly. The hotter an object gets, the wider the range of frequencies it radiates. A moderately warm object gives off only very low-frequency light, the infrared. As an object gets hotter, it starts giving off higher and higher frequency light. At first, it will give off low-frequency infrared light, then higher frequency infrared, then low-frequency visible light, which we see as red, then higher frequencies as it gets hotter.

The Earth does not get this hot, of course, so it radiates heat in the form of low-frequency infrared light. It is actually very good that the Earth starts radiating off heat as it gets warmer. If the Earth did not radiate this heat as it warms up, it would actually keep getting hotter and hotter as it absorbs sunlight, until the entire Earth melted and even vaporized. Thus, the heat radiated off the Earth back into space keeps things in balance.

Now that I have explained the basics of the physics (and hopefully not bored you too much), let's get into what all this has to do with global warming. As explained, the Earth absorbs sunlight, transfers some of this heat to the air that is in contact with the ground and water, and then radiates the rest of it back into space at a lower frequency. If there were no CO2 or other greenhouse gases in the atmosphere, almost all of the heat would radiate out into space. However, remember that CO2 and some other gases absorb light in the low infrared range, just the range that the Earth radiates out. Some of this heat energy is transferred to other air molecules, like oxygen and nitrogen, by direct contact. Some of it is radiated off again in random directions by the warmed gas. Of the radiated heat, some is radiated upward into space as it would have anyway, but some is radiated downward, where it then warms the air, water, and ground more. This is illustrated in Figure 1.1.

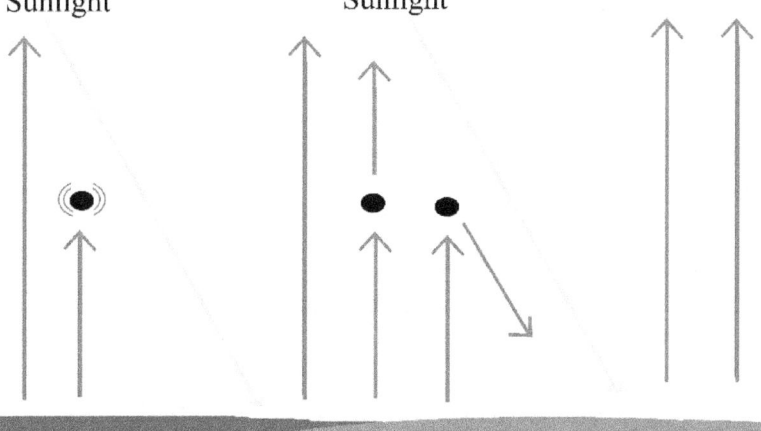

Figure 1.1

The incoming rays from the top in this picture are the sunlight, striking the land and water. The upward pointing rays are the infrared light radiated from the Earth. Some of these are escaping back into space. Some of them are being absorbed by the black dots, which are the CO2 or other greenhouse gas molecules. The molecule on the left is simply absorbing the energy and vibrating it off to other molecules. The middle one is reradiating the energy, and it happens to go upward. The molecule on the right is radiating it downward, so it is then warming the Earth again. This energy is then reabsorbed by the Earth, warming the Earth further.

This trapping of heat as it is radiated from the Earth is actually good. Without this natural trapping and returning of heat, the Earth would be about 59 degrees F (33 C) colder. The average temperature of the Earth would be about 0 F (-18 C).

The problem is that humans have been adding more CO2 by burning fossil fuels, increasing the levels from the natural levels to much higher levels. We have also been cutting down trees, sometimes burning them and sometimes using them for wood, paper, or other products. When burned, they also release CO2. Even when just cut down, they stop removing CO2 from the air.

In addition, we have also increased levels of methane, ozone, and other greenhouse gases. Methane levels have been rising partly because of leaking when we drill for oil or natural gas or while transporting or using it, partly because of release when we use fertilizer, and because of various other human activities. Ozone, nitrous oxide, and some other natural greenhouse gases are also formed by human activities. In addition, we are producing and releasing some very powerful greenhouse gases like CFCs that never existed in nature at all.

Before humans started burning fossil fuels, the level of CO2 in the air was about 280 ppm (parts per million). This means that out of every million molecules of air, 280 of them were CO2. As of the writing of this book, this has increased the level to about 408 ppm, an over 45% increase. Methane levels have increased from a natural level of under 700 ppb to about 1860 ppb today, a 260% increase. Other greenhouse gases have also increased. This means that these gases catch more of the heat. Remember the rays that made it all the way into space without hitting any greenhouse gas molecules in Figure 1.1. Now more outgoing rays are absorbed. This is portrayed in Figure 1.2.

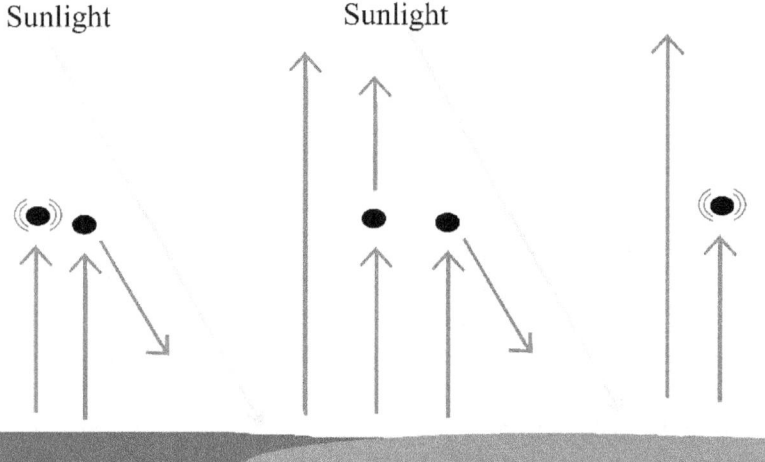

Figure 1.2

This is, of course, a very simplified description. The reradiated energy can go in any direction. It can, for example, strike another greenhouse gas molecule, which can then absorb it and repeat the process.

There are other factors that affect the temperature. For example, there are substances like sulfate aerosols in the atmosphere that reflect sunlight away from the Earth and cool the planet. These can be produced naturally, such as by volcanoes, or by humans burning dirty coal (high sulfur coal). When a lot of these are released, it can temporarily cool the planet. There are also natural variations in the sun's output that can increase or decrease the temperature. In addition, redistributions of heat within the entire Earth system (land, sea, and air), such as the El Niño allowing very warm water to remain on the surface of the ocean or changes in air patterns blowing cold air from the Arctic, can cause it to appear that the Earth has suddenly warmed or cooled, although the actual amount of heat has not actually changed. There are various other conditions that can affect the global temperature temporarily. These effects can cause confusion about whether the Earth is actually warming, which I will discuss in later chapters. However, these other factors are temporary or cyclical, while the continued increase in greenhouse gases produced by human activities is causing a continuing increase in the global temperature that can and will cause serious consequences if not stopped.

The fact that the Earth is warming does not just cause everything to get hotter. It also causes other effects, such as rising sea levels due to melting ice and expanding water, increased rain in some areas, drought in others, stronger hurricanes, and other effects on our weather. It can even cause some areas to experience unusually cold weather in brief spurts, as will be explained later. This change in weather patterns is called climate change. It is useful to understand the difference. The ongoing increase in the average global temperature is called global warming. The total effect of this warming on world weather patterns is called climate change. From this point on in this book, I will use the term climate change in most instances for brevity, unless I am talking specifically about the rise of global temperatures

Now that you have a basic understanding of how climate change works, we can get on with understanding some of the major misleading claims about it. As mentioned in the introduction, I will try to explain one false claim or misconception in each chapter of this book.

Chapter 2

The World is Not Warming

The first claim I have often heard made is that the global temperature is actually not rising. That is, that there is no global warming happening at all. Sometimes the person making the claim will cite a specific number of years, such as "There has been no global temperature rise in the last 15 years." This claim is either an outright lie or a very serious case of cherry picking. It also often makes use of the principle of making a statement that is true but leaves out certain specific facts, like why the temperature was unusually high or low in a certain year.

Figure 2.1 shows a bar graph of the annual average global combined air and sea surface temperatures for each year from 1917 to 2020. Note that the last seven years are the hottest years in over a century. In fact, the last seven years have been the hottest since humans first walked the Earth about 2.4 million years ago. So much for the world not warming.

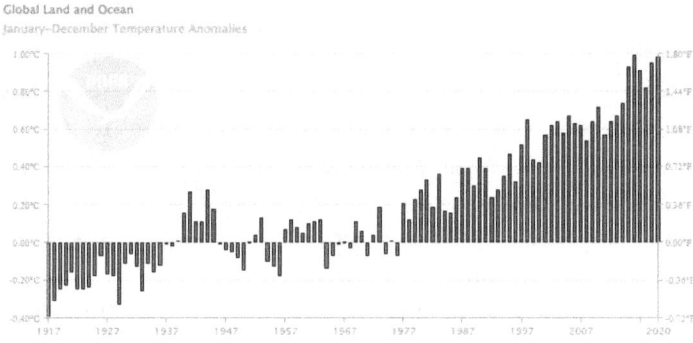

Figure 2.1

The graph is from NOAA (National Oceanic and Atmospheric Association), the US agency in charge of monitoring the world's air and water. You can use the page that generates this graph for starting and ending in any years and having a variety of conditions at https://www.ncdc.noaa.gov/cag/ . If you do use this Web page, you should pick Year-to-Date for the time scale, December for the month, Global for the region, and Land and Ocean for the region. You can select whatever start and end year you want. This will give you annual average temperatures for the entire world for those years. Of course, you can pick other settings if you are only interested in a specific region or during specific months of the year, but the settings listed above give you the big picture for global warming.

Combining global air and water temperature is the best (only, really) way to understand the global temperature, since heat energy can be transferred from air to water or vice versa, so measurements of just air or just water can give the impression of more variation than actually happens in the world. The charts on the Web site described above display the temperatures as the difference (anomaly) from the average temperature in the twentieth century. Therefore, a negative number means that the temperature is less than the average and a positive number means that the temperature is higher than the average temperature for the twentieth century. The numbers on the left side of the graph are degrees C (Celsius), and the numbers on the right side are degrees F (Fahrenheit.) If you are comparing one bar to another and trying to read the temperatures, it is important that you read the numbers on the same side of the graph, regardless of where the bar is. Otherwise, you will be comparing one number in Fahrenheit with another in Celsius.

In case you are having trouble reading the numbers on the chart in Figure 2.1, the average temperature in 1917 was 0.31°C degrees below the twentieth century average and the average temperature in 2020 was 0.98° C above the twentieth century average. That is a 1.29-degree C increase in 103 years, equal to 2.32 degrees F. That is a definite temperature increase in the last 103 years.

In case you like other time periods, the temperature increase in the 50 years from 1970 to 2020 was .92 C (1.66 F). I sometimes see the claim that the temperature has not risen in the twenty-first century. The temperature has actually increased 0.56 C (1.01 F) from 2000 to 2020. Below in figures 2.2 and 2.3 are the bar graphs of these periods from NOAA.

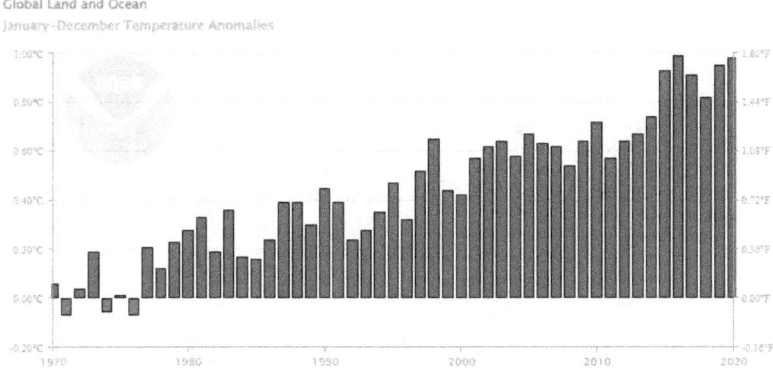

Figure 2.2

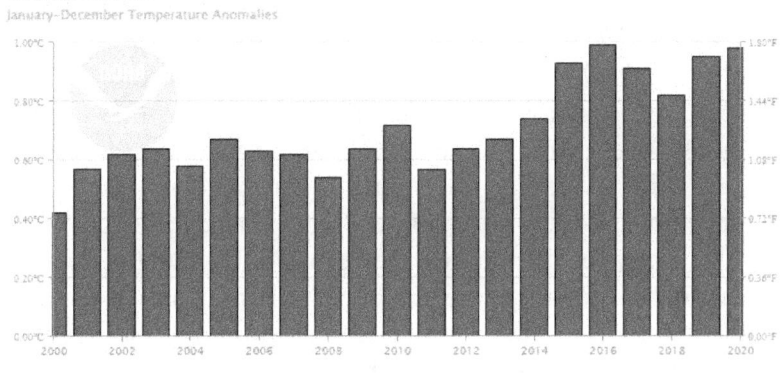

Figure 2.3

I mentioned that sometimes the claim that there is no global warming is based on cherry picking and/or leaving out important facts. The best example of this is comparing the temperature in one particularly cool recent year with a particularly hot year in the past. The most common year picked by people who are trying to be deceptive is 1998. Figure 2.4 shows a graph of 1996 through 2020.

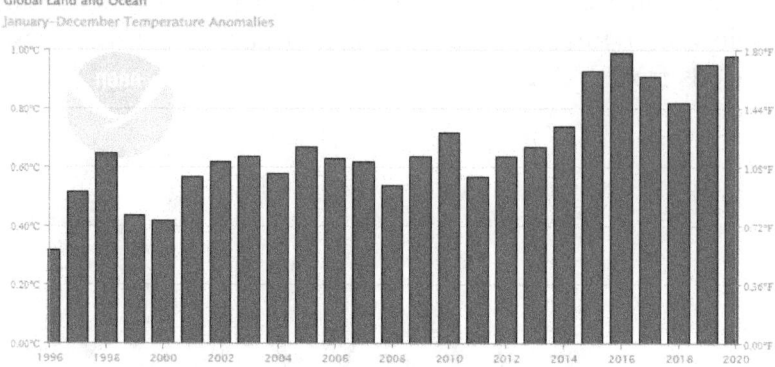

Figure 2.4

Notice the sudden short-term temperature spike in 1998. This was an El Niño year. In an El Niño year, the winds over the central and western Pacific Ocean are unusually slow. Normally, these winds churn up the water, causing the top layer warmed by the sun to mix with the cooler water below, thus cooling off the surface water. Since these winds are slow in an El Niño year, this mixing does not occur and the surface water warms up more than usual. The warmer water also warms the air. Thus, temperature readings are higher. Note that taking the entire world average, the temperature is not actually higher, because the deeper water is cooler than usual. However, since it is hard to measure the deeper water and include it in the charts, it appears that global temperatures are higher. Even if it were true that global temperatures were higher, it would still be a short-term spike in global temperatures and would not disprove the long-term warming caused by greenhouse gases. However, people who want to be deceptive about rising temperatures will often compare the measured global temperatures in years after 1998 with the measured global temperature in 1998 to make it appear that the temperature had not risen much. In 2008 and 2011, for example, it was very popular among those trying to deny global warming to say that the temperature has dropped since 1998, so the world is cooling and that proves that global warming is not happening. In making that statement, they carefully avoid mentioning that 1998 was an atypically warm year, and that the year they are comparing it to was unusually cool. They also do not mention that if you count deeper water, the temperature had actually increased since 1998.

In the next chapter, I will go into such variations in global temperatures in more detail when I discuss another deceptive statement, "The temperature is not rising steadily."

Chapter 3

The Temperature is Not Rising Steadily

I have sometimes read statements like "Scientists say that the only thing that controls the temperature is CO2. Why hasn't the temperature increased steadily as CO2 levels have gone up?" They are referring to the fact that CO2 levels have been going up steadily each year, but although temperatures have been trending upward, each year is not always hotter than the previous. This is shown in Figure 3.1.

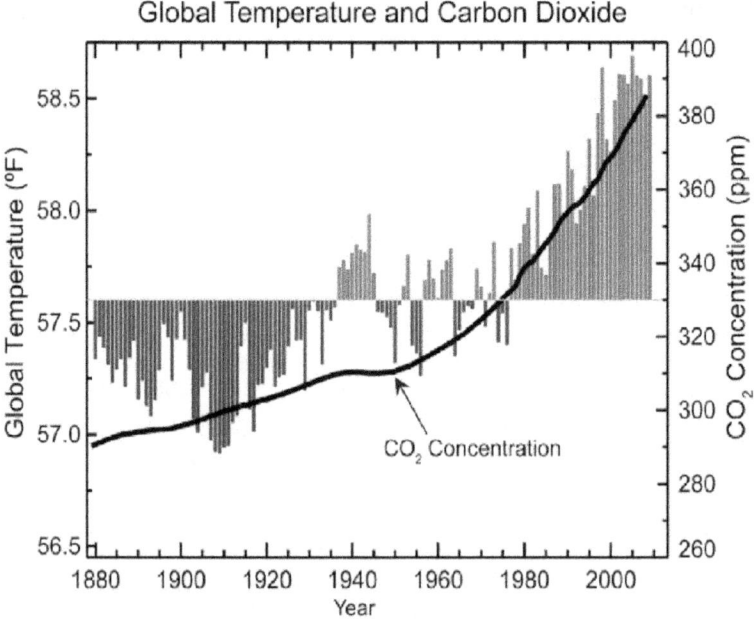

Figure 3.1

The problem with this is that they are misquoting the scientists. Climate scientists do not say that CO2 is the only thing controlling the temperature. They do say that the increase in CO2 and other greenhouse gases is the only, or at least main, cause of the current continuing

INCREASE in global temperature. That is different from saying that greenhouse gases are the only thing that affects temperature.

Remember from Chapter 1 that greenhouse gases trap heat radiated from the Earth, which originally comes from the sun. Therefore, it is more accurate to say that additional greenhouse gases have an amplifying or multiplying effect on the other factors. For example, the sun has natural cycles of about 11 years in which solar radiation increases and decreases. There are also the El Niño and La Niña periods, when surface temperatures increase and decrease. Another significant factor is sulfate aerosols, as mentioned in Chapter 1. These reflect sunlight back into space before it ever reaches the ground, causing significant cooling. Since large amounts of these sulfates are thrown into the atmosphere by volcanoes, the next few years after a major volcano eruption (or a series of coincidental smaller eruptions) can have cooler temperatures. There are other factors of this type.

You can think of these short-term fluctuations in temperature as ripples on the surface of a body of water. The ripples cause the water level to go up and down with each ripple. However, if you start raising the level of the water, the tops of the ripples will then get higher and higher and the low points of the ripples will not get as low each time you have a ripple. Looking at Figure 3.1 or 2.1 through 2.4, you can see this. You have ups and downs in annual temperature as these various conditions cause short-term ripples in the temperature, but you can see the ripples riding upward on the increasing baseline of the temperature. The highest point of the ripples in later years is higher than the highest point in earlier years, and the lowest point of the ripples in later years is still higher than the lowest point of the ripples in earlier years. If you look at Figure 3.1, you can see that the highest point in earlier years is actually lower than the lowest point in later years. For example, even the highest annual temperatures prior to

1939 are lower than the lowest annual temperatures after 1976.

A particular area of interest is the period from about 1946 to 1976. Prior to 1946, we see a fairly steady increase in temperature. Then, starting in 1946, we see a sudden drop in temperature, with wild ups and downs going above and below the twentieth century average. Then in 1977, we see above average temperatures start climbing again. What happened? Remember the sunlight blocking sulfates. During this time, there was a tremendous increase in people in various countries burning high sulfur coal, which put a lot of these sulfates into the air. This caused a drop in temperatures and fluctuations in temperature as sulfate aerosol levels fluctuated. Unfortunately, it also caused acid rain as the sulfates combined with water to form sulfuric acid, the same acid used in car batteries. This acid rain was enormously destructive, and countries started passing laws to reduce it. In 1970, the US passed the Clean Air Act, which required lower levels of sulfate pollution. This meant either switching to low sulfur coal or installing scrubbers on smokestacks. Once this law and similar laws in other countries were implemented, sulfate levels dropped. We then see the rise of global temperatures resume with increasing greenhouse gas levels.

Chapter 4

It Is All a Hoax

This is one of the most nonsensical yet ultimately effective claims made by the people who are trying to deceive people into thinking that humans are not causing climate change. It is effective because it simply shuts down the entire debate. With the mountains of evidence being collected by scientists, it gets harder and harder to claim that the facts do not prove that humans are causing the Earth to warm and the climate to change. However, if you can convince people that the facts themselves are not true, the whole argument disappears. All you have to do is convince people that all the research proving climate change is part of some vast conspiracy. What difference does it make if the charts like the ones in Chapters 2 and 3 of this book show the temperature increasing if you can convince people that the scientists are all lying about the temperatures?

There are various explanations for this mysterious conspiracy. The simplest is that all those climate scientists are faking their results to get more money for research grants. More bizarre explanations include claims that it is a plot to take over the world under some world government, a plot to destroy America for some unspecified reason, or most recently, a plot by the Chinese to ruin the American economy according to Donald Trump.

To see how ridiculous this claim is, you need to look first at the history of the science. A scientist named Eunice Newton Foote first reported her research that CO_2 is heated much more by infrared radiation than oxygen or nitrogen, the main components of air, in 1856. She conducted research by exposing glass containers containing ordinary air and containers containing CO_2 to sunlight and found that the containers of CO_2 got 20 degrees hotter than the container of air. A similar report was given by John Tyndall in 1859, in which he also demonstrated that CO_2 and other gases absorbed much more infrared light than oxygen or nitrogen using similar, but more refined and precise, measurements. Since that time, countless other researchers have repeated and refined the experiments. In 1896, the Nobel Prize winning physicist Svante Arrhenius provided mathematical equations describing just how much temperature increased with rising CO_2 levels. He was the first recorded scientist to specifically predict that rising CO_2 would raise the Earth's temperature. Of course, there were many other scientists who also contributed to this work, including verifying the results.

The point of this historical discussion is to show that the effects of greenhouse gases have been known and researched for over 150 years. This was long before there were government grants or any other incentive to do this specific type of research, so the idea that scientists would fake the results to get grants totally falls apart. It was also long before the burning of fossil fuels became a major industry, so the idea that there was some plot by the Chinese or anyone else to destroy or even hinder American development at that time makes no sense at all. At the time of the initial research, CO_2 emissions were too low to have a significant effect, so all discussion of global warming were of strictly academic interest with no economic or political impact that could have motivated anyone. It was only well after the science of global warming was well established and humans had already started to raise global temperatures that people became concerned enough to start spending money on it. There was even an episode of the popular TV show, myth busters, in which they demonstrated on screen that air container higher levels of CO_2 absorb more infrared energy.

Now let's jump ahead to more recent research. There are now tens of thousands of climate scientists studying global warming. These scientists are from almost every country on Earth. They have different political philosophies, religions, economic conditions, etc. In order for there to be a hoax, every one of those tens of thousands of climate scientists from widely different backgrounds would have to be in on this conspiracy and not blow the whistle. All of those people would have to be willing to risk ruining their careers and reputations, and possibly even be convicted of fraud and go to prison, just to get some research grants. It would be impossible to organize such a conspiracy to perpetuate a hoax of that magnitude, much less maintain it. In addition, it is not just the tens of thousands of climate scientists, it is all the support personnel. There are people who are not actual climate scientists collecting data. Most of the data come from unrelated work like routine weather monitoring, oceanographic research, and other fields. If the climate scientists were falsifying the data, all of the weather bureaus and oceanographers and routine technicians who collect data would have to be in on the hoax too, or the data would not match up. Any such conspiracy would fall apart under its own weight.

Let's look at motivation. The usual claim is that the scientists are doing it for the research grants. However, the vast majority of the money for research grants goes to equipment and other expenses. Very little actually goes into the climate scientists' pockets. Any researcher working off of government grants is basically a government employee, and government employees (other than politicians) get paid very little. Consider that to be a climate scientist, you need a college degree, almost always a Ph.D. in climate science or a closely related field. That means about 8 to 10 years of college, at great expense. Anyone with the brains and money to get through that much college could just as easily get a degree in finance, medicine, law, computer science or some other high-tech field, or any one of a dozen other high paying fields to become a lawyer, doctor, administrator, software engineer, etc. The idea that they would go into a low paying job like climatology and then engage in an illegal and immoral scam just does not make sense. This is especially true when you consider that people would still want to monitor changes in the climate, even if they were natural. After all, they would still need to know if and how the climate, regardless of why, and would therefore continue to pay climate scientists to track these changes. There is therefore no reason for scientists to attribute the warming of the world to humans if it were not true.

The other supposed motivations make even less sense. As mentioned earlier, the climate scientists come from almost every country and background, including the US. Why would every one of these scientists want to help China, or anyone else, destroy the US economy? How could all these people agree on some plan to take over the world with some totalitarian government when they have such diverse political views? Again, the idea is ludicrous when you think about it.

This does not stop the fake news from trying to spin everything to make it look like there is some conspiracy. For example, climate scientists often find new data or improved mathematical models that change the information on how much the world will warm, or even how much it has warmed. Whenever the new information indicates that global warming will be slightly less than previously calculated, the global warming deniers try to make it sound like an admission of guilt. The announcement that global warming will be less is displayed by these people with banner headlines saying, "Climate scientists admit they exaggerated global warming," making it sound like a confession of deliberate deception, rather than the announcement of a new discovery. Whenever the new data indicates that global warming is more than previously thought, they accuse the scientists of altering the data as part of a conspiracy to make global warming look worse than it is. This is the type of distortion people need to be aware of when they hear news that seems to support the conspiracy theories.

One of the best-known attempts to convince people that scientists are engaged in some kind of hoax was the so-called Climategate. In this scheme, some hackers illegally broke into computers at the Climatic Research Unit (CRU) of the University of East Anglia (UEA) in Norwich, England. They stole more than 1,000 emails and 3,000 other documents. They then sorted through these documents looking for anything to undermine the scientists' reputations. They attempted to make the emails and documents look suspicious by carefully quoting small sections, often single phrases, out of context.

One phrase quoted was "Mike's Nature trick." The use of the word "trick" sounds like they were talking about a deception, but it was really just short for "trick of the trade," a common phrase meaning a skillful technique used by specialists to solve a problem. The reference to "Nature" was that the technique was discussed in Nature magazine. In this case, the authors were referring to a mathematical technique for analyzing statistics.

Much later in the sentence, the phrase "hide the decline" was used. The hackers implied that they were discussing a decline in global temperatures, and announced that this proved that global warming was not happening at all. The truth is, they were not discussing temperature. The word "decline" in this sentence referred to a decline in northern tree growth, not temperatures. Tree growth is sometimes used to estimate temperatures when other methods of determining temperatures are not available, such as before thermometers were invented. The scientists had been graphing a variety of temperatures obtained by various methods, including tree growth, in order to show global temperatures over a long time. In the more modern times, the scientists stopped graphing the tree growth and replaced it with measurements by thermometers because these were more reliable, and because new conditions like drought and global dimming are affecting tree growth and making it a less reliable measure of temperature. They removed the declining tree growth data from the graph when they started plotting the thermometer measurements. This was what was meant by "hide the decline."

Not only did the hackers lie about what the word decline referred to, they omitted the middle of the rather long sentence. They abbreviated the sentence to "Mike's Nature trick to hide the decline." The "Mike's Nature trick" referred to something entirely different from the decline in tree growth, but by chopping out the middle of the sentence, they made it sound like a very ominous plot to conceal a decline in global temperatures.

There were other phrases taken out of context or even altered, but you get the idea. The fact that they altered the documents in this way to give the wrong impression shows the agenda of the hackers. The fact that they held the documents and then slowly started leaking them just before a major climate conference also showed that the purpose of the entire hack was to try to disrupt the climate conference. After the leak, there were six separate investigations into the Climatic Research Unit's research. One was a three-part investigation by Penn State University. Two were conducted by the University of East Anglia itself. One was conducted by the UK Parliament. One was conducted by the US National Oceanic and Atmospheric Administration Inspector General's office. One was conducted by the National Science Foundation's Inspector General's office. All of the investigations showed that the allegations that there was any faking of data were completely false. After the investigations, the Climatic Research Unit took steps to make its processes more open and transparent to prevent any future suspicions.

I should point out that the Climatic Research Unit is one tiny organization out of thousands doing research into climate change. Even if there had been any validity to the accusations of data tampering by this one group, it would certainly not prove some worldwide conspiracy by the entire scientific community. However, the climate change denying propagandists were quick to shout "Gotcha" and try to imply that this evidence was proof that all climate scientists are liars.

Chapter 5

Climate Scientists Changed "Global Warming" to "Climate Change"

As explained in Chapter 1, there is a difference between the terms global warming and climate change. Global warming refers specifically to the increase in average global temperatures, while climate change includes this plus sea level rise and all the changes in weather caused by this global warming. Scientists have used whichever term is most appropriate for the specific situation they are discussing. However, while the term global warming was used most often 40 years or so ago, people have shifted to using the term climate change as time goes by. The people who are trying to convince everyone that climate change is a hoax have repeatedly claimed that scientists made this switch as part of this vast conspiracy. The claim is that since temperatures do not go up steadily globally (see Chapter 3) and we still have cold spells locally (which I will discuss later in this book), scientists have switched to climate change as a way to avoid talking about this.

There are several flaws in this claim. The first is that climate change is not a replacement for global warming. It has a different meaning. However, the biggest deception is that it was scientists who shifted the terminology in common public use. It was, in fact, the Bush administration that carefully orchestrated the switch in terminology. President George Bush was determined to stall any effort to prevent action on climate change, at the request of the oil industry, with which Bush had close ties. The leading Republican consultant Frank Luntz conducted various public opinion studies and found that the term "climate change" was vaguer and less scary sounding to the public than the term "global warming." He therefore persuaded Bush and other Republicans, who controlled the government at that point, to always use the term climate change whenever the subject came up. With the entire government using that term, even when they were talking about actual global warming, it is no wonder the term climate change came to be used more frequently in public discussions than global warming. It is the very epitome of hypocrisy for those trying to deceive the public about climate change to manipulate the change in terminology and then turn around and try to convince people that it was the scientists who wanted to switch term.

    That said, there is a valid reason for the use of the term climate change to increase over the term global warming. When global warming first started to kick in around 100 years ago, it took a while for the secondary effects, the ones lumped into climate change, to take effect. It takes time for the glaciers to start melting and the water to start expanding to raise sea levels. It takes time for the oceans to warm up enough to increase hurricane strength. There was a lag between the global warming and the other changes in our climate. Therefore, it was natural for scientists to talk about what was immediately observable in the middle twentieth century, the warming itself. Now we are starting to see more and more of the secondary effects, the climate change. Therefore, there is more actual climate change to talk about, and more use of the correct term for these changes. Other terms have been used to try to be more precise and informative. The term inadvertent climate modification has been used, since it clearly explains that the climate is being modified, not simply changing, and that it is an accidental result of human actions. Another term that is being used more today is climate disruption, a very clear term because it indicates that the climate is being disrupted from its normal state, even from normal changes.

Chapter 6

There Is No Consensus

You may have heard that 97% of climate scientists agree that the world is warming, and this warming is caused by humans adding greenhouse gases to the atmosphere. This figure comes from a study done in 2012 by John Cook, Ph.D. In this study, he looked at 4,000 peer-reviewed research papers published over the most recent 21 years that took a stand on whether the world is warming and humans are causing it. He found that 97.1% agreed that global warming is happening and it is caused by human emissions. To make sure that he was not misunderstanding the intent of the authors, followed up with interviews of many of the authors and found the same figure.

He did this to update a similar study done by Naomi Oreskes, Ph.D. in 2004. In her article, she took 928 random papers discussing climate change. She found that 75% of the papers either specifically stated that humans were causing global warming, while 25% did not address this issue at all, but were on other subjects like research methodology or ancient natural climate change. Of the 75% that did address the subject of current climate change, all either explicitly stated that it is happening and humans are causing it, or indirectly implied agreement by evaluating impacts and/or ways to mitigate it. Absolutely none rejected the consensus that humans are causing the current climate change, or that it is happening.

There have been other surveys of this type conducted. Bart Verheggen of the Netherlands Environmental Assessment Agency surveyed 1,868 climate scientists in 2014. He found that 90% of these climate scientists with more than 10 peer-reviewed papers related to climate (as a measure of how much research they did on the subject) agreed that man-made greenhouse gases are the main cause of global warming. In 2013, James L. Powell, executive director of the National Physical Science Consortium, analyzed published research on global warming and climate change between 1991 and 2012 and found that of the 13,950 articles in peer-reviewed journals, only 24 articles disagreed with the theory that man-made greenhouse gases caused global warming. That is only 0.2% disagreement. This was a follow-up to an earlier study that looked at 2,258 peer-reviewed articles published between November 2012 and December 2013 and found only one of the 9,136 authors rejected human-caused global warming. In 2010, Anderegg, Prall, Harold, and Schneider conducted a review of publication and citation data for 1,372 climate researchers and found that 97% of the surveyed climate researchers most actively publishing in the field agree with the conclusions of the Intergovernmental Panel on Climate Change (IPCC) that humans are causing global warming. The list goes on of surveys that show an overwhelming majority of the climate scientists most actively doing research agree that the world is warming and humans are the cause. They use various methods of collecting the samples and various criteria for what constitutes agreement. They also give various figures for what percentage of climate scientists agree, ranging from the high 80% range to nearly 100%.

These researchers studying the consensus found several interesting facts. One was that as time has gone on and more and more research is done and evidence is collected, the percentage of climate scientists who agree is increasing. That is, new evidence continues to support the conclusion that humans are warming the planet.

The second fact was that the more knowledgeable a scientist was about climate science, the more convinced they are. Actual climate scientists, as opposed to scientists in other fields like geology, had a much higher percentage agreeing with the conclusion that humans are warming the Earth. Climate scientists who are actively doing research into global warming are much more likely to agree with this than inactive scientists, such as scientists who have gone into teaching instead of research. These two facts basically support the idea that the more you know about the most recent discoveries about global warming and climate change, the more likely you are to believe that humans are causing it.

The people trying to deceive you into thinking that there is no consensus generally use various tricks to give the illusion that the percentage of scientists that agree is lower. There are two basic tricks that they use.

The most common is to cite people who are not climate scientists who do not believe that humans are warming the Earth. The most famous example of this is called the "Petition Project." They claim that they have the signatures of over 31,000 scientists who do not believe that humans are causing global warming. They are careful not to say climate scientists, since that would be an outright lie, and they would be caught in that. Instead, they let people assume that the people signing the petition are climate scientists, or that scientists in any field would understand global warming. Let's look at the fact behind these "31,000 scientists."

First, all you need to be included in this group is to have a Bachelor of Science (BS) degree, the lowest level of technical education you can have. It does not have to be in climate science. In fact, it does not have to be in any science. It can be any kind of BS degree. Many of the people who signed the petition are engineers, or some other technical field such as medical doctors, computer programmers, etc. You also do not have to be a practicing scientist. That is, you do not have to have actually gone into any technical field, just gotten the BS degree. The truth is, you do not actually have to have a BS degree at all. All you have to do to be included on the list of "31,000 scientists" is to check a box that says you have a BS. There is no confirmation that you have a degree, or even that you have signed your real name. This is deliberate. The entire petition is a propaganda tool. They want as many signatures as possible, so they make it as easy as possible for anyone to add their name to the list.

It is important to understand why it is significant that you do not have to be a climate scientist to sign the petition. A medical doctor, a mechanical engineer, a computer programmer, or any other person in some other branch of science or engineering has absolutely no more understanding of global warming than the person who asks you if you want fries with your hamburger. A person who studies medicine or computer programming or mechanical engineering or any area of science other than climate science does not take courses in how the climate works. Even if you have a degree in a somewhat relevant field, like physics, oceanography, ecology, paleontology, etc., you need to understand many aspects of how the climate works to be able to understand how global warming works.

The wording of the petition is also designed to make it easy to sign it. The statement the people signing it are agreeing with says, "There is no convincing scientific evidence that human release of carbon dioxide, methane, or other greenhouse gases is causing or will, in the foreseeable future, cause catastrophic heating of the Earth's atmosphere and disruption of the Earth's climate." Note, for example, the word "catastrophic." This is deliberately designed to be an extreme position so that more people can agree with it. If you believe that humans are causing global warming and think that it will be bad, maybe even very bad, but not absolutely catastrophic, you are encouraged to sign the petition. There is also the word "convincing." This again is an absolute. If you think that there is an 80% chance, or 90% chance, or other high probability, you could still say that you are not absolutely convinced, just pretty sure.

Note that this petition is asking for people who agree with it to sign it, not people to give their opinion either way. Unlike the surveys described earlier in this chapter that took random samples of research papers or asked large numbers of climate scientists, this petition is asking you to sign it only if you agree. This is not the way you determine what percentage of people have a certain opinion. It is only the way you compile a list of people who agree with you. It is interesting to note that there are about 10 million people in the world who have the requirement necessary to be allowed to sign the petition. Finding 31,000 people out of 10,000,000 who agree with you, especially if you have no requirement that they actually know what they are talking about, is not hard. It is only 0.31% of the possible signers. Now, I am not saying that this proves that 99.69% of all the people with BS or above disagree with the petition. That is a trick used by the climate science tricksters, as I will discuss shortly. I am merely pointing out that it is not hard to get 0.31% of a group of people who have no particular knowledge of a subject to say they agree with you if you word the question in extreme terms.

Another somewhat similar tactic is to get a few very prominent scientists to say that global warming is not caused by humans. A typical example of such a scientist is Ivar Giaever, who won a Nobel prize in 1973 for his experimental discoveries regarding tunneling phenomena in superconductors. This certainly sounds impressive, but it has absolutely nothing to do with global warming or climate change. The fact that a person who knows nothing about global warming says he does not believe in it means nothing, no matter how much he knows about tunneling phenomena in superconductors. If I wanted to find prominent scientists who are not climatologists who say that global warming is catastrophic, I would have no trouble doing that either. Stephen Hawking, considered to be one of the most intelligent physicists in the world, has said that not only is global warming real, it may very well render the Earth uninhabitable. The climate change denying propagandists are strangely silent on this announcement.

Another tactic used to confuse people about the consensus is to make statements about the surveys that are misleading, or to make invalid claims about the statistical methods. One very commonly used tactic is to make the lack of an opinion sound like disagreement. For example, in the beginning of this chapter, I mentioned the study done in 2012 by John Cook. In this study, he looked at 4,000 peer-reviewed research papers published over the most recent 21 years that took a stand on whether the world is warming and humans are causing it. He took those 4,000 papers from journals that contained about 12,000 articles. The other 8,000 articles did not say that climate change was or was not caused by humans. Many of these were on entirely other aspects of climate change, like ancient natural climate change events or methodology of climate science. Since Cook was looking for what percentage of people studying current climate change believe that it is happening and caused by humans, it is natural that he would only survey the papers that took a stand on this subject. However, deceptive people will say something like "Less than 4,000 out of over 12,000 research papers published by climate scientists said that climate change is happening and is caused by humans." This is technically right, but obviously misleading. It is a bit like saying that a politician that gets elected with 97% of the votes in an election where only 10% of the registered voters voted got less than 10% of the vote, and therefore did not really win the election. When you hear figures from politicians or others saying that the consensus is only 25%, or something like that, you can be sure that they are using this tactic of counting neutral or unrelated research papers as "not agreeing with the consensus."

Chapter 7

Scientists Said the World Was Cooling in the 1970's

One statement I have heard many times by people trying to ridicule climate scientists is to say that they were all predicting an ice age in the mid twentieth century. This is usually accompanied by all sorts of suggestions that this proves that scientists are always engaging in some conspiracy to scare people, plus that they are often wrong.

The problem with this whole claim is that the basic statement is wrong to begin with. The idea that a majority, or even a significant minority, of scientists in the mid twentieth century were predicting an ice age, or even cooling, is a wild exaggeration. The scientists that even speculated that global cooling would continue were in the distinct minority, less than 10%. Unfortunately, a few newspapers grabbed onto the idea that there might be another ice age around the corner and ran sensationalist stories in the early 1970's. After all, the idea of an ice age certainly made for more exciting press than the idea of slowly rising sea levels. These stories came from several concepts.

First, as you might recall from Chapters 1 and 3, sulfate aerosols can partially block sunlight and cause cooling. Large-scale burning of high sulfur coal was common in the late 1940's through 1970's, and this did cause some cooling during this period. A few scientists speculated that if we kept dumping these aerosols into the air, we might trigger major cooling and even an ice age. The overwhelming majority of climate scientists who took a position on global temperatures accepted the idea that greenhouse gases would cause warming. What is more, once the scientists finished running their calculations, even most of the ones that had speculated about global cooling saw that they were wrong and dropped their theories. They saw that the warming effects of greenhouse gases would quickly overcome the cooling effect of aerosols. Perhaps most importantly, humanity drastically reduced its emissions of sulfate aerosols because in addition to cooling, they were causing acid rain. So, even if large numbers of scientists had accepted the idea that aerosols would cause global cooling, it would not contradict the fact that the world is now warming, because those aerosols are no longer produced in such quantities.

Another source for the stories is the fact that the world does experience periodic ice ages, and some scientists mentioned their belief that the warming trend from the last ice age had ended and we are now starting back into an ice age cycle. Some may have made comments that another ice age is coming soon. This speculation generally came from geologists studying natural ice age cycles, not climatologists studying current conditions. What many people do not understand, however, is that for geologists, an ice age coming "soon" means in a few thousand more years, not next year. A few newspapers may have run stories about this, without explaining what "soon" meant to the scientists.

One other source for the stories is the fact that at the height of the cold war, there was some concern that a global nuclear war could kick up enough dust and smoke into the air to block out the sun and create a nuclear winter. This would be like the time 65,000,000 years ago when an asteroid hit the Earth and killed the dinosaurs. The stories of this possibility also created a few stories about the possibility of a nuclear ice age.

The important point to remember is that all the stories about an impending ice age in the mid twentieth century are taken out of context or misunderstood. There never was a major acceptance among climate scientists of the idea that the world was going to have an ice age, or even start cooling. It was, at best, short-term speculation among a few scientists that has been deliberately distorted by people who want to make it seem like scientists do not know what they are talking about.

Chapter 8

Scientists Cannot Even Predict the Weather

You may have heard it a thousand times. The line generally runs something like, "If scientists cannot even predict the weather next week, how can they tell us what the climate will be a hundred years from now?" The answer is that weather is not the same as climate. It is a bit like the fact that mathematicians can say that if you toss 10,000 coins onto a table, about half the coins will be showing heads and half will be showing tails, but they cannot tell you whether a coin that lands in a particular spot on the table will come up heads or tails.

Weather is local and short-term. It is the temperature, humidity, wind speed and direction, and other things like whether it is raining in a particular place at a specific time. Calculating this is very complicated. The weather in your location tomorrow will depend on the weather today in all the surrounding areas for hundreds of miles. Every puff of wind in nearby areas may blow a little bit of hot or cold air or humidity toward our area. The exact temperature of the air and water many miles upwind of you today will determine how much water evaporates and therefore how much humidity (and therefore how much rain) you have tomorrow. For meteorologists to precisely calculate the weather at your location in the future, they would have to precisely measure the temperature, wind speed, humidity, and barometric pressure in every cubic foot of air, water, and land in all the surrounding areas, and they would have to do that over a prolonged period of time. They would then have to feed all this information into a massive computer and run models for hours to calculate how all these conditions will interact. To calculate the weather for large areas, like a state, they would have to repeat this process for each small area. When you consider the complexity of gathering all this data and then trying to compute the results in time to give you a forecast in time to act on it, you can see why weather forecasters can only give estimates and probabilities. In order to get out a forecast in any reasonable time, they need to greatly simplify the calculations and work with average temperatures, humidities, wind speeds, barometric pressure, and so on for large areas. This does not give them the precision they need for 100% accurate weather forecasts

Climate, however, is a different matter. Climate is not moment to moment or day to day, it is long-term averages. It is things like average annual temperature or average annual rainfall in an area. These normally stay fairly constant in the long run. They can vary from year to year due to influences like El Niño and La Niña, but without some external influence they will stay within a predictable range and will come back to the long-term average. As you look at larger and larger areas, it becomes even more consistent. When you look at the entire world over a long time like a year, it is much more predictable.

When it comes to global warming itself, it is even more predictable. Global temperature depends on the net input of heat, which is how much heat enters the atmosphere (in the form of sunlight) and how much of that heat radiates out. If the amount that comes in is greater than the amount that goes out, the planet will warm up. As mentioned in chapter 1, the warmer something gets, the more heat it radiates. Therefore, if more heat is coming in than going out, the temperature increases until the increase in outgoing radiation causes outgoing radiation to equal incoming radiation again. Then the warming stops. This is called reaching equilibrium. Therefore, predicting future temperature depends on calculating how much heat will be trapped by CO_2 and other greenhouse gases. This science is still not 100% understood, since there are complexities such as exactly how clouds will reflect or absorb heat, how fast ice will melt and increase absorption of sunlight, and so on, but it is certainly understood well enough to calculate whether the entire world will warm. It is therefore much easier to determine whether the world will warm up and approximately what effects this will have (such as melting ice and raising sea levels) than to calculate the exact temperature, humidity, barometric pressure, wind speed, and so on needed to predict the exact weather at a particular location.

To repeat the point of all this, climate is not weather. It is much easier to predict long-term averages (climate) than day to day events (weather).

Chapter 9

The Models Are Not Accurate

The claim is sometimes made that the mathematical models scientists use to predict climate changes are not accurate. Sometimes the claim is made that most or all of the models have overestimated how much the world would warm. That is, that predictions made some time ago said that the world would be hotter or have more severe climate change than actually has happened. The people making these claims are trying to imply that the scientists must be totally wrong about whether the world is warming at all.

The big problems with these claims is that they are simply not true in most cases. Some predictions were a little high, some a little low, but most were very accurate when you consider the number of variables involved in making the calculations. One thing that is particularly important is that there is one very good reason for some of the overestimates of global warming that has nothing to do with the basic theory. I will discuss this shortly.

One of the first projections of future warming came from John Sawyer in the UK. In a paper published in Nature in 1973, he predicted that the world in 2000 would have warmed 0.6C above the 1969 temperature. The observed temperature increase over that period was between 0.51C and 0.56C, so his calculations were only off by about 10%. That is not really that far off. More importantly, as mentioned above, there was a very good reason why his estimate was a little high. He based his calculations on the estimate that CO_2 levels would increase to as much as 400 ppm by 2000. They actually only rose to 370 ppm. It is important to understand the significance of this. His basic understanding of the mathematics of global warming were not wrong. He simply did not anticipate that humans would cut their CO2 emissions as much as they did. His error was in his prediction of the economy, not the mechanism of global warming. Of course, the people trying to discredit global warming only make the blanket statement that his model overestimated the temperature rise that would occur, deliberately ignoring how small the overestimate was and the reason for it.

This pattern is repeated in almost all of the claims that "the climate models are wrong" and therefore the science of climate change is poorly understood. In 1975, Professor Wally Broecker modeled temperature rise in the coming decades. His models were very accurate up to 2000. His estimates for global temperatures after 2000 started to be higher than observed, but again this was due to the fact that he overestimated how high CO2 levels would be, not the effects of CO2. He anticipated a level of 424 ppm by 2016, when the actual levels were 404. The IPCC models in 1990 for global warming between 1970 and 2016 overestimated global warming by about 0.15 C, which is 17%, but this too was due to overestimating the amount of CO2 that would be in the atmosphere.

Models also sometimes underestimate the amount of global warming that will occur. The 1995 estimates by the IPCC for warming by 2016 were actually about 28% too LOW. Their 2001 estimates for warming by 2016 were about 14% too low. When the people who are trying to make you think that the climate scientists are wrong talk about these, they like to say that the IPCC models are wrong, without telling you that global warming is actually WORSE than the IPCC predicted.

I will admit that there are some climate models that have over predicted global warming, as mentioned before. The 2007 estimates for 2016 were about 8% too high. Some other climate models have also overestimated the degree of global warming. There are a lot of unknowns in climate modeling. Note that these unknowns are not about fundamental facts behind global warming, like the fact that CO2 and other greenhouse gases do trap heat. Rather, they are questions of degree, such as how much heat they will trap under various conditions, how quickly the heat will be absorbed by the oceans, and so on. It is therefore important to understand that even if a climate model is off in its prediction by 8% or 10% or even more, this does not disprove the basic science. The important fact is that they all predict the world warming, and we can see that it definitely is warming at rates very close to what the climate models predict.

Chapter 10

The Climate Has Always Changed

This is probably the most repeated claim made by those who either do not understand global warming and climate change, or are trying to be deceptive. It is particularly appealing and deceptive simply because it is true. The climate has changed many times in the past. The world has periodically undergone period ice ages when ice covered about 32% of the world's land. There was even at least one time when ice covered the entire Earth. There have also been very hot periods when the area that is now the Arctic and Antarctic were like the tropics today.

The idea behind repeating over and over that the climate has always changed is to suggest that the current warming is part of some natural cycle. There is a big flaw in this suggestion, however. The climate does not simply change for no reason, any more than a rock sitting on the ground suddenly decides to roll up a hill. Every time the climate has changed in the past, there has been a specific cause, something forcing the climate to change. Scientists have studied all of the periods of climate change in the past and determined what these causes were. For example, the ice ages that occur in approximately 100,000-year cycles are caused by slight periodic changes in the Earth's rotation and its orbit around the sun. These are a natural function of the way the earth moves around the sun. Other major changes in climate have had more erratic causes. For example, there was a tremendous warming event called the Great Dying that was caused by massive volcanic eruptions that released a tremendous amount of CO_2 into the air, and also started large fires that released even more CO_2. This added CO_2 caused the world temperature to increase by about 10 C (18 F).

My point is that scientists know what caused the periods of cooling and warming in the past, and none of these natural forces are occurring now. Therefore, the current warming is not one of the natural warming cycles or other warming events. The only force that can account for the current warming is the increase in CO_2 and other greenhouse gases caused by humans.

It is especially interesting that studying these natural climate changes that occurred in the past helps prove that we are headed for severe climate change now, rather than disproving it. We can see from these previous events such as the Great Dying that an increase in CO_2, whether from natural or artificial sources, causes disastrous global warming.

Chapter 11

We Are Coming Out of an Ice Age

A related claim to the "climate is always changing" line is that "We are just still warming up from the last ice age." There are actually two variations of this claim.

The first variation of this claim relates to the natural 100,000-year ice age cycle. The claim is that the last ice age ended about 10,000 years ago, and the world is still warming up from this. This is supposed to explain the current warming. There are two really big flaws in this idea. The first is that the world has warmed up 1.15 degrees C (2.07 degrees F) in the last 100 years alone (1917 to 2017). If that had been going on for the last 10,000 since the last ice age, the temperature at the end of the last ice age would be about 223 degrees F colder than it is now. Since the current average temperature of the Earth is 61 F, that would mean that after the ice age ended, the average temperature of the earth was -162 F, which would have been lethal to just about all life on earth. Remember that I am talking about what we consider the END of the last ice age. Clearly, the Earth has not been warming at the current rate for 10,000 years. In fact, if the Earth had been warming at the current rate for even the last 2,000 years, the average temperature of the Earth 2,000 years ago would have been well below freezing 2,000 years ago. We clearly know that is not true, since our history goes back beyond that and does not describe an ice-covered planet.

The second reason we know that the current warming is not part of warming from the last ice age is that we know the warming stopped about 7,000 years ago. Figure 11.1 shows a graph from NOAA of the Earth's temperature for the last 11,300 years.

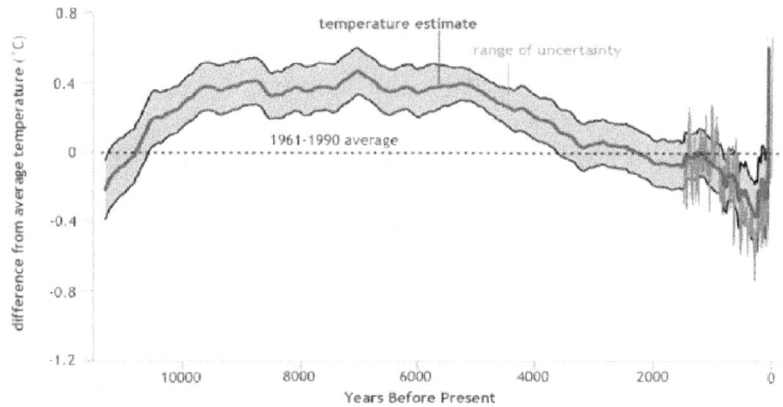

Figure 11.1

This graph shows the global annual temperature average in degrees C compared to average temperature from 1960 to 1990. Because it is hard to measure the EXACT temperature that far back, the center line shows the most likely temperature, while the top line shows the maximum possible temperature and the bottom line shows the lowest possible temperature.

Observe that the temperature is going up at a rate of about 0.15 degrees C per 1,000 years until about 9,000 years ago, then it levels off. About 7,000 years ago, it turns around and starts dropping by about 0.1 degrees C per 1,000 years. It had actually started back into a cooling cycle 7,000 years ago. At that rate, the Earth would actually be at the very beginning of a new ice age in about 15,000 years. Clearly, the world has not been in the process of warming for the last 10,000 years.

Notice, however, the extreme right edge of the graph. It shows the temperature skyrocketing upward starting about 100 years ago. The tip of this line, which represents the present, is higher than any time in the last 11,300 years. This demonstrates a sudden and dramatic move upward when the temperature had been going down slowly. This is the effect of the greenhouse gases we have been putting into the atmosphere.

The second variation of this claim that we are coming out of an ice age refers to the so-called Little Ice Age. This was a period of unusually cool weather that ran from about 1300 to about 1850, primarily in the northern hemisphere. It was caused by a combination of factors occurring at the same time, including normal solar intensity variations, periodic changes in the earth's orbit, unusually high volcanic activity producing large amounts of sunlight blocking aerosols, changes in ocean currents that carried cooler water into these areas, and a few other minor factors. The coincidental occurrence of several cooling forces at the same time produced this cool period.

This "Little Ice Age," which was not really an ice age at all but simply a period of cool weather, ended around 1850. At that time, the temperature had returned to the temperature at the beginning of the Little Ice Age. Any further warming would not be coming out of that "ice age," since we were already out of it. To put it another way, the temperature now is far higher than at the beginning of the period. If we were coming out of an unusually cool period, the temperature would have stabilized at the temperature it was at the beginning of that period, not skyrocketed since then. In addition, the temperature did not continually rise since the end of that period, which it would have done if this were a continuation of the warming coming out of that period. In fact, the global temperature dropped slightly after we came out of the Little Ice Age. Global warming actually started kicking in around 1917, when the CO2 levels really started to climb rapidly as a result of industrialization. Therefore, the current global warming is not the result of some natural trend continuing from the Little Ice Age.

Chapter 12

It's The Sun

The cry of "It's the sun" is very popular with people who deny that humans are causing the increase in the Earth's temperature. Their claim is that the sun is the only thing that affects the earth's temperature. They claim that any change in the Earth's temperature is due to variations in the sun's output. In order to support this claim, they point to the up and down variations that occur in annual average temperature, which are often caused by variations in the sun's output. They point to the cyclical variations in the sun's out, which do occur. These are not regular cycles, but do occur in cycles that range from 9 to 14 years, averaging 11.1 years. There is also a claim that there is a 400-year cycle, but this has absolutely no basis in science. In order to establish that such a cycle exists, we would have to have observed solar output for an absolute minimum of several cycles over 800 years, and several times that to really confirm it. Since we have only had even rudimentary observations of solar activity since about 1612, just over 300 years ago, any claim that there is evidence of a 400-year cycle is pure nonsense, and seems to be made up out of thin air.

People who claim that the sun alone controls the temperature also often claim that climate scientists say that CO_2 is the ONLY regulator of global temperature. This is not true. Anyone who says that climate scientists say that CO_2 alone controls the temperature is either mistaken or is being deliberately deceptive. It is important to understand that CO_2 and other greenhouse gases trap heat, they do not generate it. This means that greenhouse gases amplify or multiply the effect of the sun. Consider a car sitting in the sun. If the windows are closed, the car will get much hotter inside than if the windows are open. The glass windows let sunlight in but trap the heat. However, the temperature of the car still depends on how brightly the sun is shining on the car. In the same way, the presence of man-made greenhouse gases traps whatever energy we get from the sun, but you still get variations in temperature from year to year as the sun and other factors vary. The average temperature is just hotter.

Getting back to the basic claim that changes in solar output are responsible for global warming, there are many ways to refute these. I could probably write a whole book just discussing the reasons why these are nonsense. However, to name a few:

Solar intensity is not increasing, at least not significantly. Figure 12.1 shows a chart of both solar irradiance and global temperature from 1880 to the present. The darker top line is a running 11-year average of the Earth's temperature. The lighter line superimposed on it is the yearly temperature. The bottom darker line is a running 11-year average of solar irradiance. The lighter line superimposed on that is the yearly solar irradiance.

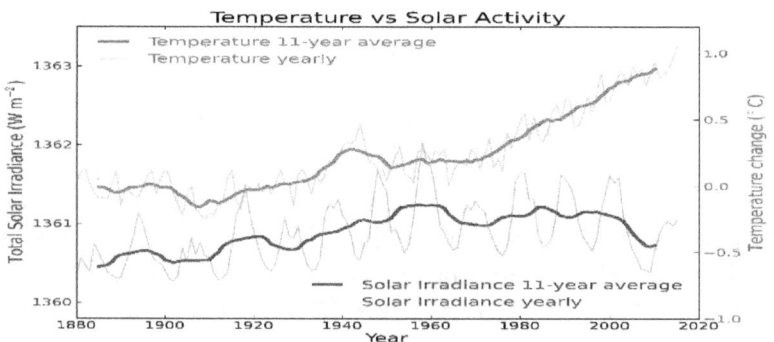

Figure 12.1

Notice that there is a slight correlation between the lines at the beginning, but this disappears in the mid 1950's. In fact, solar irradiance starts dropping in the late 1050's, but global temperature continues upward. Clearly, even if there was an increase in solar irradiance and it did contribute to global warming, the effect of increasing greenhouse gases is so much stronger that when the solar irradiance drops off, the increased greenhouse gases overwhelm the effect of decreasing solar irradiance.

The above discussion was about solar irradiance, the aspect of solar output most likely to affect the temperature. However, people who claim the sun is the only thing that affects the temperature usually talk above sunspots and general solar activity. Figure 12.2 shows a comparison of sunspots and global temperatures from 1978 to 2008. (I apologize for not being about to find a more recent chart that makes this comparison.) The top line shows global temperature, with the steadier line being multi-year averages and the more erratic being annual. The bottom line shows solar energy output varying with sunspot activity, again with a steadier line being averaged and the more erratic line being yearly.

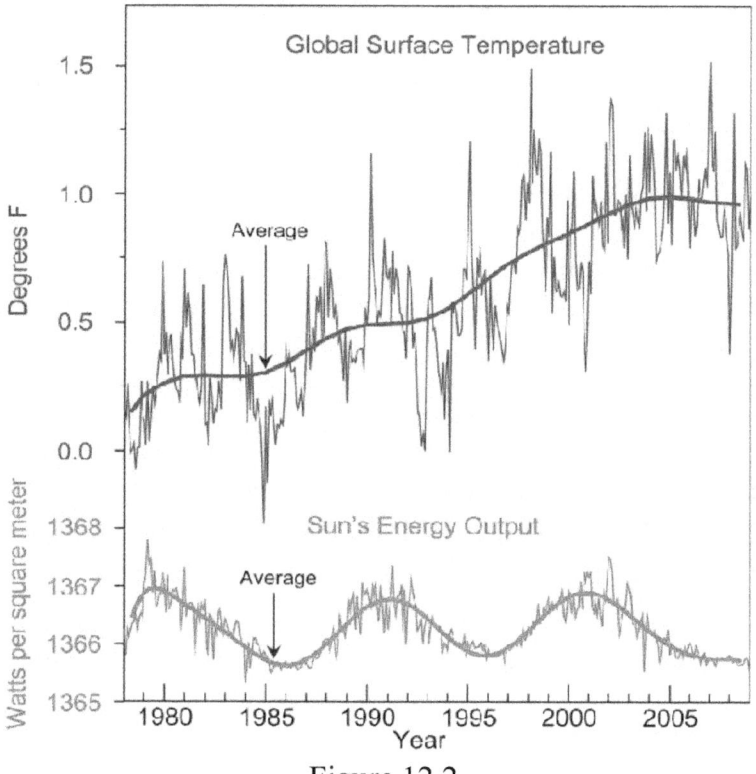

Figure 12.2

You can see that although the overall sunspot activity has not changed in the long term, the temperature has climbed. Again, clearly the sunspot cycle is not driving the upward movement of the global temperature.

There are several other interesting pieces of evidence that show that the increase in temperature is not caused by an increase in solar activity. Satellite data shows that the upper atmosphere is actually cooling. If the solar output were increasing and driving an increase in the atmospheric temperature, the temperature would be increasing throughout the atmosphere. On the other hand, the effect of the greenhouse gases is to trap heat in the lower atmosphere. Since less infrared radiation from the ground is reaching the upper atmosphere, it makes sense that the upper atmosphere would be cooling.

Another piece of evidence is that the greatest increase in temperature is seen at night, not in the daytime. Both the daytime and nighttime temperatures have been increasing, but the difference between nighttime temperatures now compared to 100 years ago is greater than the difference between daytime temperatures now and then. If the temperature were increasing due to more solar energy, you would expect the daytime temperature while the sun is shining to increase more, while the nighttime temperature would not be as affected. However, when you consider that the effect of the greenhouse gases is to hold onto energy that would normally escape into space, it makes sense that nighttime temperatures would see the greatest increase. Before the increase in greenhouse gases, the heat escaped more quickly into space at night. Now that the greenhouse gases are capturing, holding, and then more slowly releasing the heat energy, the nights are staying warmer longer.

Chapter 13

CO2 Lags Temperature

By studying both the current increase in global temperature as CO2 rises and prehistoric events where temperature has increased with increased CO2, we can see a clear correlation between CO2 and temperature. When CO2 goes up, temperature goes up. However, some people have suggested that the CO2 went up because the temperature went up, not the other way around. That is, they say that the temperature went up first, and the increase in CO2 lags behind this increase in temperature.

This comes under the category of a statement that can be true, but the people that say it are leaving out an important piece of information. It is true that increasing global temperature can cause an increase in CO2. There are many reasons for this. Warm water cannot hold as much CO2 as cold water. The world's oceans contain a great deal of dissolved CO2. If an external force causes the ocean temperatures to increase, they will release some of that CO2, causing an increase in atmospheric CO2. There are other effects. Increased temperature can cause an increase in growth and activity of some soil bacteria that release CO2 from the soil. Increased temperatures can increase the rate of plant decay. Increased temperatures also lead to more forest fires. There have been times in the past when an increase in temperature caused an increase in CO2 in the atmosphere.

The flaw in the claim that this disproves man-made global warming is that it assumes incorrectly that just because A causes B, B cannot cause A. They are implying that somehow the fact that warming can cause an increase in CO2 disproves the idea that CO2 can cause warming. This is a major logical fallacy. It would be a bit like saying that if John and Bill get into a fist fight, the fact that John punched Bill first proves that Bill never hit John.

The truth is that warming can cause an increase in CO_2 and increasing CO_2 can cause warming. This is what is called a positive feedback effect. The increasing CO_2 increases the temperature which increases the CO_2 which increases the temperature, and so on until all the CO_2 that can be released has been or something forces the temperature down. The cycle can start with either the temperature increasing or the CO_2 increasing, but either way, it builds on itself.

This is what causes the periodic ice ages we have seen in the past. There are slight periodic changes in the shape of the Earth's orbit, the Earth's tilt, and the Earth's wobble. When all of these changes combine just right, they cause the Earth, especially the northern hemisphere, to get cooler. This causes the oceans to absorb more CO_2, which causes the world to cool more, which causes more CO_2 to be absorbed, and so on. It also causes more ice to form, which reflects more sunlight, which causes more cooling, which causes more CO_2 to be absorbed and more ice to form, and so on. The world then follows this downward spiral into an ice age, when ice covers much of the world. The reverse process causes the ice ages to end. When the periodic changes in the shape of the Earth's orbit, the Earth's tilt, and the Earth's wobble cause the world to warm slightly, the oceans start releasing some of the CO_2, which causes the world to warm, which causes more CO_2 to be released from the oceans, and so on. Melting ice also feeds this cycle since less heat is reflected back into space.

The fact that the end of the ice ages sometimes starts with warming, even though it is fed by the release of CO_2, is the basis for the claim that warming causes CO_2 to increase, not the other way around. They ignore the back-and-forth cycle between rising CO_2 and temperature increase, and just focus on the fact that CO_2 started to increase after the temperature went up. They also ignore the times in the past where the cycle was started by the release of CO_2 from some source, such as a massive volcano. The best example of this happening was the Great Dying extinction event 252 million years ago, discussed previously in chapter 10. In this case a super volcano eruption released massive amounts of CO_2 into the atmosphere. This raised the Earth's temperature about 10 C (18 F). Clearly, this was a case of CO_2 causing an increase in temperature, not the other way around.

Chapter 14

It's Cold in (fill in the blank)

You may often hear claims that because we had a cold day or week somewhere sometime, the world must not be warming. Often you will hear cries of "It's cold outside. Where is all that global warming?" Senator James Inhofe, a rather notorious climate change denier, is famous for throwing a snowball on the Senate floor on a cold winter day to mock the idea that the world is getting warmer.

It is easy to pick isolated events and create an intuitive feeling that they prove something. I could point to the fact that there has been no terrorism in my small town, so maybe there is no such thing as global terrorism. However, isolated events do not prove or disprove an overall trend.

First, remember that there is a difference between climate and weather. Climate is long-term conditions over a fairly large area. Weather is very short-term conditions in a small area. Global warming or climate change does not mean that every square mile of the planet changes steadily and consistently. There will always still be variations in the weather, and sometimes these local short-term variations will be larger than the overall trend. If the temperature of an area naturally varied from 20 F to 90 F before global warming set in and the global temperature average has gone up by about 2.07 F, that would mean that the new temperature range is 22.07 F to 92.07 F, and 22.07 F is still cold enough to get snow. Anyone who suggests that throwing a snowball proves that the world is not warming is either very confused or playing you for a fool.

Even in a warming world, you will still have some very cold, even record cold days. However, one interesting metric of global warming is to look at how many record-low temperatures you have around the world compared to how many record-high temperatures you have. It is significant to note that the number, duration, and intensity of record highs each year is now significantly more than the number of record lows. In a world where the global temperature stayed the same, you would expect the number of record lows to be the same as the number of record highs. This is not the case. In the 1950s. there were 1.09 times as many record highs as record lows. In the 1990's, there were 1.36 times as many record highs as lows. Between 2000 and 2010, there were twice as many record highs as record lows. As of December 5 of 2017 (most recent date I have records for), there were 2.36 times as many record highs as record lows in the previous 365 days. The fact that we are having many more record highs than lows over time indicates that the baseline around which temperature swings is centered is rising.

Another significant fact is that global warming can even cause short-term extreme cold conditions locally by temporarily shifting weather patterns. In particular, the loss of sea ice in the Arctic is making the polar vortex and jet stream unstable. The polar vortex is a zone of frigid air that encircles the Arctic. Normally, it keeps the frigid air in the Arctic. However, rising temperatures are changing wind patterns, causing frigid air to break out of this normally contained zone and come south. This can cause severely cold short-term winter weather in northern North America, Europe, and Asia. Those who would want you to think that global warming is a myth or hoax gleefully point to these isolated occurrences to convince you that the world is not warming. It is worth noting, however, that when this happens, warmer air from the south moves up into the Arctic to replace the cold air that has moved south, giving these areas record warm periods. Therefore, global temperatures have not dropped, just been redistributed.

Bottom line: Please do not be fooled by isolated, short-term instances of cold weather. The global temperatures are rising.

Chapter 15

Earth Has Had Lower Temperatures with Higher CO2

Some people still claim that CO2 does not really trap heat and raise the temperature, despite over 150 years of experimental proof. One fact that seems to support this is that there have been times in the past when CO2 was actually higher than it is today, but the global temperature was as cool or cooler than it is today. How could that be if CO2 controls temperature?

The answer is to remember that CO2 is not the only factor controlling temperature. People who are trying to confuse the public on this issue often say that climate scientists say that CO2 is the only factor controlling temperature, but that statement is simply not true. Climate scientists understand that CO2 amplifies the effect of the sun by trapping heat that would escape without that CO2 in the atmosphere, but temperature is still affected by many other factors.

The first factor is how much solar energy comes to the Earth in the first place. The sun has been very slowly warming up at a rate of about 1% every 100,000,000 (100 million) years. Before going on, I want to point out that this does not explain the current warming. The rate the sun is warming up accounts for a warming of a degree or so every 100 million years, not 1.15 degrees C in the last 100 years. However, this slow warming does explain why the Earth could be cooler than it is now at times in the distant past when CO2 levels were higher.

A person who did not believe that humans cause global warming recently pointed out to me that during an ice age 450 million years ago, the CO2 level was 11 times what it is today. At that point, the sun was about 4%

dimmer. Naturally, the lower level of energy from the sun had a lower heating effect, which tended to counter the effects of the higher CO2 and allow the Earth to be cooler than today. In addition, there were differences in the shape of the land masses that contributed to cooling by allowing ice to form more easily and reflect sunlight.

There are other factors. Our information on the levels of CO2 hundreds of millions of years ago is very incomplete. Some evidence that appears to indicate high CO2 levels may have actually just been short-term spikes in CO2 levels in the middle of major ice ages. These spikes did not last long enough to melt the ice, so the ice ages did not end. Scientists therefore give a range of possible CO2 levels. Of course, people who are trying to convince you that CO2 does not cause warming usually give the higher levels when talking about CO2 levels during cold periods.

Chapter 16

Antarctica Is Gaining Ice

The claim that Antarctica is gaining ice rather than losing it has been cited as evidence that the world is not warming. This is a typical example of taking one fact out of context and deliberately drawing an incorrect conclusion from it.

First, only some parts of Antarctica are gaining ice. Other parts of Antarctica are losing ice from increased melting due to global warming.

More importantly, the reason that some parts are gaining ice is due to global warming. Antarctica gets its ice by snow falling on it. The snow comes from water condensing out of the atmosphere, and the water in the atmosphere comes from evaporation from the ocean. Because the oceans have been getting warmer due to global warming, there has been more evaporation. More evaporation means more water vapor to condense as it moves over cooler areas. This means more precipitation in the form of snow in Antarctica. The fact that this increased participation in Antarctica is coming down as snow does not prove the world is not warming. The average temperature in most of Antarctica has been well below freezing. Even though there has been considerable warming in Antarctica, it started out cold enough that most of it still averages below freezing. When the increased precipitation from warmer oceans comes down in these areas, you get snow which adds to the ice.

Another fact is that as some of the ice has been increasingly melting, the ice has been spreading out. For example, as ice melts during warmer weather, fresh water from the melting ice flows outward into the ocean. As the weather cools again in fall and winter, this fresh water freezes into ice. This spreading ice results in the area covered by ice in the sea increasing, even if the total volume is not increasing. You also have calving glaciers breaking up into smaller pieces, which spreads ice. Much of this ice falls into the ocean as icebergs, creating a larger surface area of ice even though there is actually less cubic volume of ice. I have had people try to tell me that this increased ice area proves that Antarctica is cooling, rather than warming. It proves nothing of the sort, of course. The fact that gigantic ice formations are crumbling in the heat and spreading out certainly does not prove that it is getting colder. It demonstrates the opposite.

A quick update. As of February, 2018, the latest satellite pictures show Antarctic sea ice at the lowest on record. The effect of the warming oceans is now overcoming even the effect of the breaking up and spreading of the ice from land caused by warmer air temperatures.

Chapter 17

Sea Levels Are Not Rising

I have mentioned several times so far that one effect of global warming is rising sea levels. However, some people are now trying to claim that sea levels are not rising at all, or that to the rising is slowing down or stopping. They try to support these claims by quoting some reports that seem to support their claims. However, they achieve this by cherry picking reports of specific locations or time periods.

Regarding location, sea level rise is not the same in every location on Earth. It also depends on what you are measuring sea level relative to. Various areas of land are themselves rising or sinking due to various geological forces. If you are measuring sea level relative to a shore that is on land that is rising faster than the sea, it can seem in that area that sea level is actually dropping. On the other hand, if the land is sinking or being eroded in a particular area, sea level can appear to be rising even faster than the global average. People who deny that sea level is rising can easily pick a specific location where the land is rising so that sea levels do not appear to be rising and say, "Sea levels have not risen at all in this area." To determine global sea level, you need to average sea level rise over many areas. This should include the middle of the ocean. Satellites can measure sea level in the middle of the ocean, where it is not relative to a specific area of land. This gives a valuable reference.

Based on all measurements, NASA and NOAA have determined that average rate of sea level rise for the entire world is currently about 0.125 inch (3.18 mm) per year. This rate is a considerable increase over the rate before 1990, which was about 0.043 inches (1.1 mm) per year. That is nearly triple the rate from 30 years ago. The rate is expected to continue to increase rapidly. Considering the rising temperature, this makes sense. As temperatures all over the Earth rise, more and more ice-covered land that previously had an average temperature below freezing now has an average temperature above freezing. Therefore, more and more ice is melting. In addition, ice that has recently started melting will melt faster as the temperature continues to rise. The glaciers are gigantic, so it can take centuries for one to melt completely. That means that as the area where glaciers are melting expands and gets hotter, the rate that water is being added to the ocean will increase, increasing the rate of sea level rise.

Now let's look at sea level rise over time. I mentioned that people who deny that sea levels are rising cherry pick dates as well as locations. This is because sea level rise is not constant. Sea levels can even appear to fall for short periods. Figure 17.1 is a graph of sea levels.

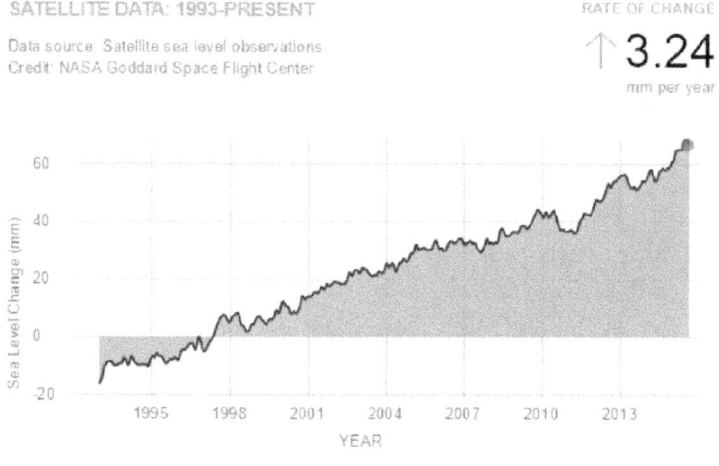

Figure 17.1

You can see that although there is a clear overall pattern of sea level rising, there are brief periods when sea levels fall. How can this happen? Remember that we are talking about water actually in the ocean. Sometimes weather conditions temporarily transport water from the oceans to land. During La Niña, rainfall increases dramatically in northern South America, Southeast Asia, and Australia. This means that water that has evaporated from the oceans and would normally fall back into the ocean instead falls far inland. It can take from a few months to over a year for this water to flow back into the ocean. See https://www.scientificamerican.com/article/a-scientist-explains-the-mystery-of-recent-sea-level-drop/ and http://onlinelibrary.wiley.com/doi/10.1029/2012GL053055/full. During this time, ocean levels will be lower. Just as people who deny that the temperature is rising will carefully pick two years where the second year was cooler than the first year, people who deny that the sea level is rising will carefully pick two periods of time where sea levels are lower in the second period than the first. It is important to look at long periods of years or decades to see that sea levels are rising.

Update notice: The latest report from NOAA lists 2.4 meters, about 8.2 feet, as the worst-case scenario for sea level rise by 2100. Although this is an upper range estimate, the report suggests that recent studies showing a higher-than-expected probability of massive glacier collapse in Antarctica significantly increases the probability of such catastrophic sea level rise. https://tidesandcurrents.noaa.gov/publications/techrpt83_Global_and_Regional_SLR_Scenarios_for_the_US_final.pdf

Chapter 18

Melting Ice in Water Does Not Raise Water Levels

Along the lines of the claims that sea level is not rising discussed in the last chapter, I have encountered the argument that global warming cannot raise sea level because ice cubes melting in a glass do not raise the water level. That is, since ice that is already in water does not raise the water level when it melts, melting icebergs cannot raise sea level. Therefore, global warming cannot raise sea levels.

The first part of this is true. Melting ice that is floating in water, including icebergs, does not raise the water level. The reason for this is simple. Ice is about 11% larger than the liquid water than formed the ice is. Water is very unusual in that when it freezes, it forms a solid crystalline structure that is larger than the liquid form. Very few other liquids do this. Almost all other liquids get smaller when they solidify, not bigger. Because the solid form of water (ice) is larger than the liquid form, it floats. About 11% of the ice is above the water it is floating in, with about 89% below. When the ice melts, the 11% above the water is added to the water, which would raise the level, but the part below shrinks by 11%, which lowers the water level by exactly the same amount. Hence, no change in water level.

The second part of the statement, that this proves that global warming would not raise sea level, misses an important point. As explained in previous chapters, it is not the ice that is floating in the water (icebergs) that raises sea levels. It is the ice that is on land, glaciers and snow which melts and flows into the oceans that raises the sea levels. In addition, when glaciers melt, some of the water from the melting ice flows down through cracks in the ice and gets below the glacier. This forms a layer of water below the glacier that can allow the glacier to slide down the land into the ocean. Having a glacier slide from land into the ocean has the same effect as if the entire glacier had melted and flowed into the ocean. That is, it will cause sea levels to rise, only much more suddenly than is the glacier had slowly melted.

Another source of sea level rise is that as the water in the ocean warms up, it expands, as almost every other substance does when it gets warmer. As the water expands, sea level naturally rises.

Thus, it is true that floating ice does not raise water levels and therefore melting icebergs do not raise sea levels. This has nothing to do with the fact that global warming does cause sea levels to rise by melting glaciers, causing glaciers to fall into the ocean, and expanding water through warming.

Chapter 19

CO2 is a Trace Gas

I have heard the argument that CO2 cannot possibly cause global warming because it is a trace gas, a gas with very low concentrations in the atmosphere. This is basically a gut-feeling argument with no basis in scientific fact at all. Although it may seem intuitively that a tiny amount of something can have no effect, the truth is that very small amounts of many substances can and do have large effects. The gas perfluoroisobutene is toxic as levels of 1.2 ppm. CO2 levels in the air are currently over 333 times that level. To put it another way, perfluoroisobutene gas can kill you at levels of less than 0.3% of the current level of CO2. Arsenic in drinking water is considered unsafe at 10 parts per billion, which is 0.01 ppm. That is less than 0.003% of the level of CO2 in the air. You can smell (Z)-8-tetradecenal at a concentration of 0.009 ppb (parts per BILLION) in water. Of, course, the warming effect of CO2 is not exactly the same thing as toxicity or detectability. However, the idea that CO2 cannot affect escape of radiant energy from the Earth because its concentrations are so small is purely an intuitive argument with no basis in fact at all. Showing that small concentrations of substances can have large effects does demonstrate how wrong our intuition can be.

To understand the greenhouse effect of CO2, consider the fact that if there were no CO2 in the air at all, the average global temperature of the Earth would be -18 C (0 F). The current global average temperature is about 15 C (59 F). (That may seem colder than you would expect, but bear in mind that this is including the polar regions.) That is, reducing the CO2 level from about 405 ppm to 0 ppm would cool the Earth by about 33 C (59 F). Clearly, even small amounts of CO_2 in the atmosphere can have large effects. Remember that other greenhouse gases released by humans, such as methane and CFC's, are much more powerful greenhouse gases than CO_2, and therefore can have considerable effect at even lower concentrations.

Chapter 20

It Is the Water, Not the CO2

I have heard the argument that CO2 does not control global temperatures, water vapor does, because water vapor is a much more powerful greenhouse gas than CO2. While it is true that water vapor is a more powerful greenhouse gas than CO2, this argument misses an important point.

This point is that we are concerned about the INCREASE in the global temperature. Only greenhouse gases that have been recently increasing are causing an increase in temperature. Therefore, water vapor could only account for the increase in temperature if there had been a steady increase in water vapor over the last 100 years due to natural causes. There has been no mysterious force of natural force causing more water to accumulate in the atmosphere.

It is worth noting that there is, however, an unnatural cause of an increase in water vapor: global warming. As CO2 levels have gone up, the temperature of the water and air have both increased. Higher water temperature leads to more evaporation into the air. In addition, hotter air can hold more water vapor, resulting in more water vapor in the air even at the same relative humidity. This additional water vapor in the air then adds to the increase in global warming. Thus, water vapor does add to global warming, but only because the global warming had already been started by increased CO2. This is called a positive feedback loop, and is one of the most dangerous aspects of global warming because it can multiply the effect of adding CO2.

Chapter 21

Humans are Not the Main Contributors of CO_2

If they give up on trying to convince people that CO_2 is not the cause of global warming, the people who are trying to confuse you about the cause of global warming will sometimes suggest that humans are not responsible for the increase in CO_2. They suggest that there are natural sources of CO_2 that generate much more CO_2 than humans.

This argument falls flat on its face for several reasons, but the most obvious is that CO_2 levels have suddenly increased from about 280 ppm to about 408 ppm since humans began their massive burning of fossil fuels. Prior to that, CO_2 levels have remained between 180 and 300 ppm for the last half a million years, with levels around 280 for the last 10,000 years or so. If CO_2 levels were increasing due to natural forces, why did this increase so coincidentally start just as humans started adding CO_2 to the air?

The simple truth is that natural CO_2 levels have been in balance for a very long time. Plants breathe in CO_2 and exhale oxygen. Animals eat plants, inhale oxygen, and generate CO_2. There is some absorption of CO_2 by water, and some release of CO_2 by water. Some rocks absorb CO_2, and some natural processes like volcanoes release it. The amount of CO_2 released overall by natural forces has obviously been balanced by the amount absorbed for a very long time. I say obviously, because if it were not balanced, we would either have almost no CO_2 in the air if it were being absorbed more than emitted, or massive amounts of CO_2 if more were being emitted than absorbed. CO_2 levels are currently increasing by about 3 ppm per year. Since the current level of CO2 is about 408 ppm, if natural forces have been increasing CO2 levels at that rate, that would mean that CO2 levels 136 years ago would have been 0.

So where do people get this idea that natural forces contribute more CO2 than humans? This is one of the worst cases I have ever seen of stating something that is true while leaving out an important piece of information. It comes from looking only at the generation of CO2 by any natural force and not at the absorption by that same force. For example, I have actually heard the claim that plants emit more CO2 than humans. Huh? As I recall my high school biology, plants absorb CO2 and release oxygen. So where do we get the ridiculous claim that plants release more CO2 than humans? Well, part of it is technically true. Plants absorb CO2 from the air and use it to produce cellulose (the material they are actually made of), starches, and sugars. They then burn a very small amount of that sugar to power their own metabolic processes, so plants do re-release a very small percent of the CO2 that they absorb. They also release CO2 as they decay after they die. Humans release about 41 billion tons of CO2 into the atmosphere every year by burning fossil fuels. All the plants in the world, living and dead, release about 439 billion tons of CO2 into the air, so technically, plants do produce more CO2 than humans if you only look at what they emit. However, plants also absorb 450 billion tons of CO2 from the air to produce that sugar, starch, and cellulose. Somehow, the people who are trying to get you hopelessly confused about the contribution people are making to global warming manage to forget about the 450 billion tons of CO2 absorbed by plants and just tell you that plants exhale 439 billion tons of CO2 while humans only produce 41 billion tons per year by burning fossil fuels.

I have also heard the claim that the oceans emit more CO2 than humans. Again, partially true. The oceans absorb CO2 and also release some of it. This is just the normal process of any gas going into and out of a liquid that it is in contact with. Normally, the waters of the Earth release about 332 billion tons of CO_2 per year, but they also absorb about the same amount. This is a natural equilibrium. The fossil fuel companies are happy to tell you that the oceans release 332 billion tons of CO_2 per year while humans only release 41 billion tons, without mentioning that the same oceans reabsorb those same 332 billion tons of CO_2. In fact, because humans have been increasing the CO_2 in the atmosphere, the equilibrium has been disturbed. The oceans are now absorbing more CO_2 than they release, about 6 billion tons more, for a total of 338. This is good for the atmosphere and helps slow down global warming, but unfortunately the CO_2 has been combining with H_2O to form the acid H_2CO_3, which is making the oceans acidic and harming marine life. This can be almost as big a problem as the warming itself. It is particularly harmful to shellfish, like shrimp, crabs, lobster, oysters, and clams. Industries that harvest these are already seeing a marked decline. Even more important is that organisms that form the bottom of the food chain, like phytoplankton and zooplankton, could die out. If that happens, the entire ecosystem of life in the oceans could collapse, resulting in the loss of a major human food source.

So, next time someone tells you that plants, the oceans, or any other natural force is emitting more CO_2 than humans, you know they are trying to make a fool of you. All the other forces are in balance, absorbing the same amount of CO_2 as they release. Only humans, by digging up long buried carbon sources like coal, oil, and natural gas and burning it, are producing more CO_2 than they absorb.

Chapter 22

CO2 is Plant Food

The claim that CO2 is plant food, suggesting that the more CO2 the better, seems to have caught on, especially with politicians. The claim is that the more CO2 we put into the air, the more plants, especially food crops, will grow. I have even seen the suggestion that farmers should be required to pay fossil fuel power plants for the CO2 they are producing.

As in many cases discussed in this book, there a grain of truth in this statement. Plants do use CO2 to build their own structure, as explained previously in this book. Under some conditions, increasing CO2 levels in the atmosphere can increase plant growth. This is sometimes done in greenhouses or other controlled plant growing situations to generate greater plant growth.

However, there are two problems with this claim. First, it assumes that the benefits will continue to increase as the levels of CO2 go up. That is a bit like suggesting that since people need about 2,000 to 2,500 calories a day for energy, it would be enormously healthy to force feed a person 10,000 calories a day. Second, it focuses very tightly on one possible benefit of higher CO2 levels without looking at any of the downsides.

The problem with the idea that more CO2 is always better for plants is that CO2 is not the only thing that plants need. They also need water, nitrogen, phosphorus, potassium, magnesium, sulfur, and calcium. They also need a certain amount of sunlight for every molecule of CO2 they use. Once they have used up all of any of the things they need, they cannot make use of additional CO2. Whatever they run out of first becomes the limiting factor in the plant's growth.

The most common limiting factor is water. Since plants combine CO2 with water to form cellulose, additional CO2 is useless if the plant runs out of water. This is particularly significant since one of the effects of increased CO2 is hotter, drier weather which dries out the soil and limits plant growth. Thus, while CO2 is providing one nutrient, it is removing a more important one. This is important because there is always CO2 in the air, even without humans burning fossil fuels. However, there is not always enough water for plants to survive.

In addition, heat directly affects the ability of plants to thrive, even if they have enough water. It is significant to note that the more valuable plants, food crops, are more complex than simpler forms of vegetation and therefore more vulnerable to climate disruption. For example, corn, soybean, rice, and cotton show a dramatic drop in growth when the temperature is above 32 C (89 F). Corn is particularly susceptible, showing an average drop of 7.4% in yields for every degree C that the Earth warms. Another interesting example is the maple trees of Vermont that provide the famous Vermont maple syrup. These trees are declining as temperature rises because they are not adapted to the heat.

An additional problem is that warmer temperatures are allowing the spread of parasitic insects and diseases that are harmful to plants into regions that were formerly too cold for them. The plants in these regions tend to have no immunity to these invaders because they evolved without having to deal with them until now. Prominent examples include (but are not limited to) the southern pine beetle and Colorado potato beetle, as well as various fungi. In some areas of Colorado, 80% of the existing forests have been killed by the pine beetle. The dead trees are a forest fire hazard for tens of thousands of people.

The fossil fuel industries and their proxies like to point out recent NASA reports that some areas of the world have been "greening," showing an increase in green vegetation, due partially to increased CO2. What they do not mention is that the vast majority of this greening is moss and weeds, simple plant life that can easily grow under adverse conditions. In fact, the increase in weeds is actually creating problems for farmers.

There are also possible secondary effects of the heat. For example, if pollinators like bees and butterflies die out due to climate change, we will see massive losses of crops. Even if the pollinators or other symbiotic forms of life that plants depend on do not die out completely, climate change can disrupt the relationship between the plants and other species. For example, warmer weather has been found to cause some plants to bloom too early, before migrating pollinators reach the area. Rising temperatures also enable the spread of insect pests that destroy plants. Also, severe weather like storms and flash floods are becoming more frequent and severe due to climate change, and these damage crops. Many fields of crops have been destroyed by sudden flooding or windstorms.

There are other adverse effects, although not as significant as an actual loss of food crops. There is growing evidence that when unusually high amounts of CO_2 do increase plant growth, it does it by enabling them to produce more sugars at the expense of using important nutrients like zinc and iron. The resulting foods are more sugary but less nutritious for humans. In addition, recent studies show that plants exposed to increased CO_2 are less able to utilize nitrogen, causing plants to have less protein. This is not just a matter of plants producing more sugar and therefore having less protein proportional to the sugar. It has been found that the excess CO_2 actually inhibits the ability of plants to absorb or use nitrogen, so even if the size or sugar content of the plants does not increase, protein content goes down. Researchers found that even adding more nitrogen to the fertilizer does not help increase nitrogen absorption (https://phys.org/news/2015-06-carbon-dioxide-air-restrict-ability.html). To put it simply, excess CO_2 is turning some of our healthier foods like wheat, rice, peas, and soybeans into junk food. In addition, Researchers at the University of Gothenburg have conducted experiments that show that at higher levels of CO2, plants lose the ability to absorb nitrogen, even if nitrogen levels in the soil are boosted. This results in lower protein content in plants, making them less nutritious.

Bottom line: While increased CO2 may help some plants under some conditions, the beneficial effects of this added CO2 are very limited and eventually become harmful, while the harmful effects of the changing climate are getting worse as CO2 levels are going up.

Chapter 23

Global Warming Will Be Good

Just in case they cannot convince people that global warming is not happening, or that it is not caused by humans, the fossil fuel industry has a backup strategy. They and their proxies are now claiming that global warming will be good. They are claiming that we will all be better off in a warmer world. After all, who does not like to be toasty warm? They list a variety of supposed benefits. They claim that plants will grow better in warmer weather, that there will be fewer weather-related deaths, that new land will be made available as cold areas warm up and the ice melts, more rain, and lower energy costs. Let's look at these claims.

We have already looked at the claim that plants will grow better with more CO2 in the previous chapter. This is slightly different, because now they are suggesting that warmer temperatures themselves will cause plants to grow faster. As usual, there is some truth to this. Sometimes cold weather will reduce the growth of plants. However, as explained in the previous chapter, weather that is too warm can also reduce crop yields. The question is, is the current temperature of our farmland too low, too high, or just right. The answer is that the current temperature is just about right, and any increase will start to lower crop yields. When you think about it, this is very logical. Between natural selection and deliberate breeding, farmers have been developing crops that will grow best under the conditions they have been growing their crops for hundreds of years. If an ear of corn or a stalk of wheat is not suited for the current climate, it will not grow well. The farmer will not plant seeds from that specimen. Any strains of any crops not suited to the climate farmers are currently growing their crops in have been weeded out of the growing process through many generations of farming. The crops have settled into their environment. Any change in the environment will be bad, not good. The maple trees of Vermont are a perfect example, in that we are already seeing a dramatic drop in syrup yield that is directly attributed to the warmer climate. This does not even consider indirect effects of higher temperatures, such as drought, spread of pests as the climate warms, and more severe weather effects like tornadoes attributed to climate change. Bottom line: Any change in an environment that plants have spent hundreds or even thousands of years adapting to will almost certainly be hard on the plants.

What about weather-related deaths? They claim that more people die of cold than heat, so warming up will reduce deaths. However, determining how many people die of cold vs. heat is complex. Are we talking only about actually freezing to death or dying of heat stroke? It should be easier to avoid dying of cold than heat. If it is too cold you can wear heavier clothes, start a fire, or various other ways to get warm. If it is too hot, there is not much you can do about it unless you have access to air conditioning, which many people in less developed countries do not. What about indirect causes of death? Do auto accidents caused by ice on the road count as a death caused by cold? Does carbon monoxide poisoning from a gas heater run indoors during a cold snap count? Does starvation or death by thirst count as death by heat? What about deaths from diseases caused by the spread of disease to areas that were previously too cold for the diseases to live there? How about deaths caused by increased strength of hurricanes, or flash floods driven by climate change? Deciding what deaths to attribute to climate change is a big judgment call. The World Health Organization predicts that climate change will cause about 250,000 deaths per years between 2030 and 2050, but they are counting both direct deaths from heat and all of these secondary effects. However, a recent study calculated that unless greenhouse gas emissions are reduced dramatically very soon, by 2100 75% of the world's population will be exposed to lethal heat waves for 20 or more days per year. (https://news.nationalgeographic.com/2017/06/heatwaves-climate-change-global-warming/) This does not guarantee that they will all die, of course. Some may have access to air conditioning or other means of cooling off. However, this does put people at risk of dying from a simple power failure, which becomes more common when utilities are strained, and many people will not have access to these modern conveniences in poorer countries. In addition, some people must go outside for prolonged periods to work or for other reasons, exposing them to the extreme heat.

There is one other problem with the theory that global warming will reduce deaths from cold. As mentioned earlier in this book, global warming has been disrupting the polar vortex, resulting in extreme cold spells. When this happens, people have been exposed to colder temperatures than they are used to. This is expected to increase as the polar vortex becomes even more unstable. Thus, even if you assume that cold is more dangerous than heat, climate change is still a major threat. It can actually create very extreme cold spells in areas that are not used to them, the most dangerous kind of cold. As an interesting coincidence, as I write this, the northern US is experiencing record-breaking cold weather.

Bottom line: Climate change is likely to increase deaths, not decrease them.

What about the possibility that warmer weather will make cold areas more livable, thus opening up a new frontier of land? As cold areas of land warm up, there may be some places that become suitable for settling. However, the question is, will this offset the land lost? As just mentioned, up to 75% of the world's population is living in areas that are expected to experience killer heat waves by 2100. If these areas become basically uninhabitable in the long run because of these heat waves, we are looking at a lot of land lost. Then there is the problem of land lost due to sea level rise. Because access to the ocean for shipping and fishing is so useful, 75% of the world's major cities are by the sea, and about 40% of the world's population lives within about 37 miles of the coastline. Naturally, this means that a lot of very expensive infrastructure is located there. Various studies have estimated the cost of sea level rise to be between $100 billion and $1 trillion per year by 2050, depending partly on whether you consider increased hurricane damage due to higher sea levels. Considering the fact that any new land opened up by warmer weather will be barren wasteland at latitudes that get less sunlight then the currently settled land, it would be very hard to figure out a way that new land opened up by melting ice or land made more pleasant by warmer temperatures in northern Canada or Siberia could come close to offsetting, much less surpassing, the value of land that will be lost due to unbearable heat or sea level rise.

One other surprising fact is that often even the areas you would think would welcome a warming climate suffer because of it. In Alaska, for example, most of the ground has been frozen solid mud (called permafrost) for as long as people have inhabited the area. In fact, about 25% of the northern hemisphere is on permafrost. This forms a hard base that people have constructed buildings and roads on. Now that permafrost is beginning to melt, causing buildings to collapse and roads to warp as the ground gives way. This is causing major damage to the infrastructure of civilizations built in these areas. Another problem is that much of the coastline in Alaska and other northern areas is protected from waves by sea ice. As the sea ice has been melting, these areas have seen increasing erosion of inhabited land on the ocean.

Bottom line: While some land may become more useful as the world warms, it looks very unlikely that this land will actually be nearly as valuable as the currently usable land lost to rising seas, desertification, and other disasters caused by climate change.

What about the idea that climate change will bring more rain, which might help crops grow or produce other beneficial effects? It is true that climate change does increase rain in some areas, but it tends to do this in areas that already get plenty of rain. The increased rain results in too much rain, causing flooding. North Carolina, for example, has recently experienced "Rain Bombs," intense rainfalls often of 10" or more in a less than an hour. Rain is caused when water evaporates in a warmer area and the water vapor is blown into a cooler area, where the water vapor condenses into water and falls as rain or snow. Hotter air can hold more water. This means that when the air cools, there is a lot more water to condense, and it comes down in strong downpours instead of a gentle rain. The result is that we are seeing more flash flooding. At the same time, in hot areas where there tends to be a lot of evaporation, the greater heat causes more evaporation, causing drought. Thus, as far as water is concerned, global water tends to take from the poor (the drier areas) and give to the rich (areas with plenty, maybe too much, water).

What about the idea that rising temperatures will lower energy costs by reducing extreme cold temperatures? Such claims make no mention of the increased costs of air conditioning caused by extreme heat, the energy cost of transporting water as droughts increase, pumping water from increasingly flooded areas like Miami, and other needs for increased energy as the climate changes. Considering the many ways the climate is changing, it is hard to say whether there will be a net increase or decrease in energy use worldwide. Whatever the net energy use change is, it will probably be small either way, and certainly not enough to compensate for the massive other expenses caused by climate change.

Chapter 24

A Few Degrees Will Not Make a Difference

Many times, I have heard a refrain something like, "Global warming is supposed to raise the temperature two degrees. Who cares? The temperature varies by more than that every day." Sounds reasonable at first glance. However, there are many flaws in that argument.

First, the estimate that global warming will raise the temperature only two degrees or so by 2100 is a very conservative estimate and is essentially obsolete. Newer studies place a very real possibility of much higher levels. A British investment firm, Schroders (hardly the type of organization you would expect to be pushing for environmental controls), has recently conducted a study that indicates that the temperature could go up by 7.8 C (14 F) by 2100 if we continue releasing CO_2 at our current rate. While this is the highest estimate I have seen from a reputable source, estimates like a 6 C (10.8 F) increase by 2100 are not uncommon from major sources like NASA and PricewaterhouseCoopers (PwC), a multinational professional services network headquartered in London. Scientists have described a 6 C increase in temperature as a doomsday scenario that the world cannot adapt to. Note that these figures are based on the assumption that we do not take action to significantly reduce our greenhouse emissions. Figures like a 4 C rise are considered likely if we make significant reductions.

Second, consider that the figures, whether they be 7.8 C or 6 C or 4 C or even 2 C, are global averages for the increase. Some areas are heating up faster than others. If, for example, the average increase in global temperature is 2 degrees and half the Earth only heats up by 1 degree, this means that the other half of the Earth heats up by 3 degrees. Of course, that is just an example to demonstrate the point. In some areas the air temperature may not rise much at all, while in others it heats up considerably. Unfortunately, the areas that are heating up fastest, at least when you are talking about air temperatures, are the areas that matter the most. Because water can absorb a great deal of heat energy from the air while only warming a much smaller amount itself, air over the open ocean is not warming as fast as air over the land. Therefore, land (where most people live) takes a greater proportion of the rise in air temperature than the seas. Air over the Arctic is also warming twice as fast as the global average. This is bad because it is speeding up the melting of ice. As the white ice melts, sunlight strikes the darker sea water or land, resulting in more warming. This is one of the feedback effects. It is also responsible for the destabilization of the polar vortex, as explained in other chapters. Bottom line: Just because global averages only go up a few degrees does not mean that the places people live will.

Third, when you say that normal daily or seasonal temperature fluctuations are greater than the increase in global temperatures, remember that the increase is added to the daily or seasonal fluctuations. If, for example, the temperature would normally range from 70 F to 94 F on a given day and global warming causes an increase of 6 degrees F, the range has changed to 76 F to 100 F. Thus, the global warming has increased the chances of severe heat waves. Such severe heat can not only affect people, but it can also affect crops, infrastructure, and other things important to people. In some areas, for example, recent heat waves have actually cause asphalt roads to melt. Likewise, there are areas where the temperature naturally sometimes goes beneficially low, killing mosquitoes, pine bark beetles, and other pests. By shifting the temperature range upward a few degrees, it allows these pests to spread to new areas, increasing the chance of disease and other damage.

Fourth, and this is very important, there is a difference between short-term changes in temperature and sustained changes. For example, a short period during the day of temperatures above freezing will not permanently melt glaciers. Any ice that melts will refreeze when the temperature drops again. However, if the long-term average temperature goes above the melting point, you see major ice melting and sea level rise. In addition, sea level rise is also caused by warming sea water expanding. Since water takes a long time to heat up, it is the long-term average temperature that causes water to heat up and expand, not daily or even seasonal fluctuations. The same is true of other conditions. A prolonged higher temperature evaporates water from soil and water supplies more than a short spike in temperature, leading to drought. Recent studies, such as one done by Dr. Manoj Joshi, professor of climate dynamics at the University of East Anglia, indicate that a mere 1.5 C increase in global temperatures will result in 25% of the Earth becoming much drier, approaching desert like conditions.
(http://www.independent.co.uk/environment/global-warming-world-land-arid-desertification-climate-change-study-a8139896.html). Long-term warming is causing more evaporation off the oceans, making hurricanes worse. Bottom line: A permanent change in the temperature can and does produce negative effects that short swings of temperature do not.

Chapter 25

The World Survived Global Warming Before

I have heard comments like, "The world has had periods of global warming before and survived. Why should we worry about it this time?" Note that this is different from the idea that because the world has warmed before naturally and therefore all global warming is natural discussed in Chapter 10. These people are saying that we should not be concerned about man-made global warming because the world has survived natural warming before.

What this argument misses is that when it happened before, it was pretty terrible. The best example is the Permian–Triassic extinction event which happened 252 million years ago, which most scientific evidence indicates was caused by a massive release of CO_2 from a super volcano in the region that is now Siberia. This period is referred to as "The Great Dying" because so much of the life on Earth was wiped out. Up to 96% of all marine species and 70% of terrestrial vertebrate species becoming extinct. Not only individual species went extinct, but about 57% of all biological families and 83% of all genera became extinct. This was partly due to the heat, and partly due to the increased acidity of the seas because of the CO_2. Much of the planet became basically uninhabitable. It would be a terrible loss to our world if that happened again.

The other problem is our civilization. If that type of warming happens today, it will be especially hard on humans because we are largely dependent on a stable infrastructure. We have cities near the shore that will be flooded by rising seas. The latest report from NOAA has a high-level projection of 2.5 meters (over 8 feet) of sea level rise by 2100 (https://tidesandcurrents.noaa.gov/publications/techrpt83_Global_and_Regional_SLR_Scenarios_for_the_US_final.pdf). Our society is very interdependent. Food and other necessary goods are shipped from all over the world through ports that can be submerged by rising seas and destroyed by super powerful hurricanes, transported by rivers that can dry up, and taken on roads that have been known to literally melt or buckle due to extreme heat. Our lives depend on power that is generated by a steady flow of water, both to turn turbines in dams and to cool power stations like nuclear and even fossil fuel power. Having a relatively sudden change in our environment would be so disruptive to our civilization that our economy and social order could very well collapse.

Chapter 26

Stopping Global Warming Would Destroy the Economy

I have heard the claim that if we try to stop or reduce global warming by switching to renewable energy or other methods, it will be so expensive that it will bankrupt the economy. I have seen claims that it would cost $100 trillion dollars by 2100 to even try to slow down global warming. These figures are very suspicious, especially since the reports that claim such figures usually contain comments like "if in fact climate is controlled by carbon dioxide" and "even if we actually faced a climate catastrophe" and "the political hot air and ever-larger government subsidies of today's inefficient green technologies" and "President Obama's very ambitious rhetoric." The reports that quote such huge figures then go on to say that spending all that money will actually have almost no effect on global warming at all. In short, articles containing such claims clearly come from the same people who have a clear political agenda and are throwing out every possible argument why we should not fight global warming.

Such arguments, as usual, are made by cherry picking figures and ignoring many others. They assume very high costs for solar panels and windmills. They also ignore savings that come with these sources of energy.

The basic idea behind such claims that switching to clean energy would be very expensive assumes that solar panels, windmills, and other sources of renewable energy are very expensive. Such claims are way out of date. It is true that when they were first developed, these sources were very expensive and awkward, just like any new technology. However, as the technology has progressed, the cost has dropped at an incredible rate, somewhat like the price of computers and other high technology. The cost per watt of silicon solar cells has dropped from $76.00 in 1977 to about $0.30 today. It is now less than 0.4% of what it cost in 1977, and the price drop is continuing. Industry experts are saying that the price of solar power is expected to drop another 50% by 2020. The price drop is partly due to advances in technology, and partly due to economics of scale. Economics of scale means simply that the more of something you manufacture, the cheaper each individual item becomes because the cost of some of the basic manufacturing equipment is shared by more items. Note: These costs have been expressed in terms of cents per watt put out by the solar cell. Considering the life expectancy of solar cells and other factors, this comes out to about $0.10 per kWh today, and about $0.05 or less if the expected halving occurs by 2020.

It is important to note that this is just the drop in price of silicon PV cells. This is one of the oldest and most established types of solar cells. Other types of solar cells are being developed that use other materials and hold the potential to be even cheaper than silicon cells. Some possibilities include perovskite and organic materials. It is likely that each type of solar cell will be cheapest under different circumstances. For example, one type might be cheaper per watt but be less efficient, producing less power per unit of solar cell area. Such solar cells would be economical in areas where large amounts of land are cheap and available. On the other hand, another type of solar cell might be more efficient, producing more power per unit of solar cell area. Such solar cells would be the best type to use in circumstances where there is limited area to place them, like some roofs. Some solar cells are transparent and can be placed over windows. There is even technology being developed that allows you to simply paint solar material directly only a wall or roof like applying a coat of paint, allowing almost any surface to become a solar cell. Thus, as the technology improves, more and more options are becoming available, allowing people to use the cheapest option in any situation, bringing costs down much further.

Another type of solar energy is concentrated solar thermal power, commonly referred to as CSP. In this system, many mirrors focus sunlight onto one area, which naturally gets extremely hot. This heat is used to drive a generator, just like when you burn coal, oil, or gas, but without releasing any CO_2 or other pollution. This has the advantage that the systems can easily store the heat in molten salt. It is easy to construct the system to hold enough molten salt to store about 10 or even 15 hours of heat energy. This allows the system to run 24 hours a day, eliminating the objection that solar power does not provide power at night. While this type of solar has been more expensive than solar cells that convert sunlight directly into electricity, the prices on this type of solar are also dropping rapidly. In fact, in Spain the price that developers were bidding for new installations dropped in half from May to October of 2017. The new price was $0.05 per kilowatt hour, making it less than half the price of coal. Of course, this may be under ideal conditions. However, other estimates are around $0.06 per kilowatt hour by 2020. It is estimated that for every doubling of the amount of concentrated solar power installed, the price will drop by about 30% due to economics of scale.

The cost of wind energy is also dropping rapidly. In 1980, the cost was about $0.63 per kWh. Today, the cost can be less than $0.02 per kWh. In Colorado, a utility company that solicited bids for a wind power plant received multiple bids at $0.0181 per kilowatt hour. Even more exciting news was that even when they included battery storage so the wind farm could supply power when the wind is not blowing, the price was still $0.021, just over 2 cents, per kilowatt hour. That is considerably less than coal and very competitive with natural gas. The cost of building a wind farm is now actually cheaper in many areas than continuing to run an existing coal plant. Be clear about that. It is actually cheaper today in many areas to shut down a coal plant that you have already built and build a wind farm than to keep supplying the coal plant with coal. Building a new coal plant under these circumstances make no economic sense at all.

In addition to solar and wind generated electricity, there are other sources of renewable energy. These include geothermal (heat from within the Earth), wave power, tidal power, deep ocean currents, hydropower, and solar water heating. Most of these have more limited application than solar and wind because they are only useful in certain areas, such as on seashores or in areas where the Earth's heat comes close to the surface. However, the technology to use these is being developed and they can contribute to the entire renewable energy portfolio.

The price of renewable energy, especially solar and wind, will continue to drop. There are three reasons for the drop.

1) Continued improvements in the technology of the devices themselves.
2) Continued improvements in the technology of the manufacturing methods of the devices that bring down manufacturing costs.
3) Economics of scale. As explained previously, the more of something you manufacture, the lower cost per unit unless there is a shortage of raw materials. Estimates of price drop for renewable energy sources for every doubling of the amount produced are estimated to be 35% for solar cells, 30% for concentrated solar thermal power, 21% for land-based wind, and 14% for offshore wind. This creates a feedback loop where the more of any type of renewable energy is installed, the lower the cost, which increases the demand for construction, which lowers prices more, and so on.

I have mentioned that some forms of renewable energy, such as concentrated solar thermal power, basically have built-in storage. Storing energy from solar, wind, and some other forms of renewable energy has always been a problem because such sources are not as constant and controllable as burning fossil fuels. Critics of solar and wind energy have always made a big deal of this, even claiming that renewable energy is a danger to the power grid because of fluctuations. There has been some truth to this in the past. However, this situation has rapidly changed and is continuing to change. Improvements in battery technology have made it practical to store large amounts of power at reasonable costs. Batteries and other forms of energy storage such as pumped water and thermal storage are now being used to provide steady electrical current for renewable energy sources. For example, in Australia, Elon Musk just completed and put online a 100-megawatt battery backup for the grid that maintains power 24/7 year-round. Adding battery backup to grid level renewable energy adds as little as 1/3 cent per kilowatt hour to the price of power. It is interesting to note that renewable energy with stored energy can be even smoother than fossil fuel power. Fossil fuel plants generate power by heating water until it boils and then using the steam to drive generators. It takes a bit of time to drive up the heat if demand for power increases suddenly, just as it takes time for water to boil when you put a pot on the stove. With batteries, power output increases almost instantly when demand increases.

For comparison, the price of fossil fuels is between $0.05 to $0.17 per kWh. Contrary to claims that renewable energy cannot compete without government subsidies, it is actually fossil fuel companies that would go out of business without government help. An October 2017 report by Oil Change International reveals that fossil fuels receive over $20 billion in fossil fuel subsidies per year from US federal and state governments. The World Bank determined in 2016 that the world's oil and gas companies receive $5 trillion annually in government subsidies. Even with this, the fossil fuel companies are running scared. Trump is reported to be considering a plan to actually pay coal power plants $15 per ton to burn coal, because coal cannot compete with renewable energy. In some states, politicians who receive large contributions from fossil fuel companies have passed laws requiring people who install solar panels on their own roofs to pay the power companies monthly fees when they do not buy power from the utility. In some states, they are even trying to outright ban utilities from using solar and wind power. After years of complaining about government subsidies for renewable energy, the fossil fuel industry is now demanding that the government prop them up.

Electric power consumption is not the only source of CO_2 emissions, of course. In the US, transportation is a larger contributor than power production. This problem can be solved by switching to electric cars that are charged from renewable energy sources. Critics of electric cars have made three basic complains:
1) They do not have enough range (travel distance between recharging).
2) They take too long to recharge.
3) They cost too much

While these used to be serious problems, technology is once again solving them.

The range of many modern electric cars is more than adequate. Some of the top-of-the-line models (like Tesla) get over 300 miles per charge, many makes and models get over 200 miles, many more get well over 100 miles, and almost all get over 80 miles. That is more than enough for most needs such as commuting to work. They are especially efficient in city and other congested driving because they consume no fuel when not actually moving.

Of course, range can be a problem if you like to go on driving trips. The solution to this is plug-in hybrids. These can run off of batteries until they batteries run out, then switch to the gas engine. These have range from about 14 miles to 53 miles (2017 Chevrolet Volt) on the battery before the as engine needs to kick in.

The charging time problem is closely related to the driving range problem. After all, if charging time is low enough and there are enough conveniently located charging stations, the electric car becomes equivalent to the gas-powered car in convenience. The number of charging stations is rapidly increasing. These are being installed by the power companies (e.g., California's Pacific Gas and Electric is about to install 7.500 in California), the electric car companies (especially Tesla), and the owners of the destinations themselves for the convenience of their customers and employees. There are currently about 16,000 public charging stations in the US, with a total of about 44,000 outlets. Of course, the main location for charging cars is at home with your own charging station, but being able to charge up your car while at work, at a hotel, or at a restaurant certainly adds to the convenience.

As for charging speed, although it takes about 75 minutes for the Tesla supercharger to charge a car from 0 to 100%, it can charge a car to 50% in about 20 minutes. Thus, you can easily top it off during a stop for food or while at work. Many other cars have similar rates of about 30 minutes for 80% using fast chargers, which are usually available at public places, but at home chargers take longer to charging times, such as 4 to 13 hours for the really slow low power charger models.

As for cost of electric cars, the same forces that are reducing the cost of renewable energy are working on electric cars. Technological improvements in batteries, both the batteries themselves and the manufacturing processes, are causing prices to drop. The Smart electric car sells for around $24,000, the Ford Focus and Hyundai Ioniq electric for just under $30,000, the Volkswagen E-Golf and Nissan Leaf for just over $30,000. Several others sell in the $32,000 range. The Chevrolet Bolt EV, with a range of 238 miles and a fast charge speed of 30 minutes, sells for under $37,000. These prices put electric cars in the range of gas-powered cars. Note that these prices are not counting any government subsidies or tax breaks.

When thinking about cost, it is important to understand that electric cars are much less expensive to use once you own them. The cost of electricity per mile is a fraction of the cost of gasoline. Depending on local electric and gasoline costs, it costs about $3.50 to go 100 miles in an electric car, compared to about $8.00 for gasoline. Maintenance costs are much lower too, due to the simplicity of electric cars. There is no oil to change, many have no transmission fluids, no alternators and belts to wear out, etc. There are simply fewer moving parts to wear out.

It is important to note that batteries for both electric cars and storage for wind and solar farms are being improved rapidly. There is evolutionary improvement in the existing batteries. There are also totally new types of batteries are being researched that could bring revolutionary increases in storage capacity, reduced charging time, and decreased price. An auto company called Fisker Inc. claims to be a few years away from selling a solid-state battery that will cost less than half as much as today's batteries, give a range of 500 miles (804 kilometers), and charge in minutes. They also will not have the fire hazard of lithium batteries. Toyota is also experimenting with solid state batteries. An Israeli company called StoreDot has announced a new type of battery that they say can be charged enough to drive an electric car 300 miles in 5 minutes. The Mercedes car company has invested in this company and plans the use the battery in their cars. Other car and truck companies are also planning to use this battery. The Massachusetts Institute of Technology chemist Donald Sadoway is pioneering liquid metal and metal mesh batteries. Toshiba has announced a major improvement in lithium batteries that they say will double the range of the car and allow it to be charged in 6 minutes.

It is clear that renewable energy and electric cars will eventually replace fossil fuel for purely economic reasons alone, even without any attempt by the government to encourage it. The question is, will it come in time to prevent catastrophic global warming and climate change? I have mentioned positive feedback in previous chapters. If the temperature rises too much before we stop adding greenhouse gases to the atmosphere, these feedback loops will all kick in and the temperature will continue rising even after we stop adding greenhouse gases. It is therefore imperative that we reduce our greenhouse emissions, especially CO_2, as quickly as possible.

What can be done? Remember that once solar and wind becomes more common, the price will drop to well below the price of fossil fuels, and energy will become cheaper and cheaper due to economics of scale. This will benefit the economy tremendously. It is just a matter of tipping the balance slightly toward renewables as quickly as possible. There are many practical actions the government can do that will not have a harmful effect on the economy. Here is a partial list of possible options.

1) Eliminate subsidies for fossil fuels, and preferably transfer these subsidies to renewable energy sources. This will make renewables less expensive and fossil fuels more expensive. Power companies will build more renewable energy plants and fewer fossil fuel plants. Since renewable energy is already very close to the cost of fossil fuels and often cheaper, this will not raise the price of energy significantly, and any increase will be very short-term and will be replaced by a significant price drop within a few years.

2) Create a carbon tax, where fossil fuel producers or sellers must pay a small tax (although I prefer the term environmental impact fee) for each ton of CO_2 their products will produce when burned. The income from this can be used to subsidize clean energy and electric cars. Again, the price of dirty energy will go up, but the cost of clean energy will go down. This will encourage people to use clean energy.

3) Mandate clean energy by requiring that power companies get a certain percentage of their power from clean sources and car companies produce a

certain percentage of electric vehicles. By being guaranteed a certain number of sales, clean energy companies will be encouraged to produce a certain number of units, such as solar panels or windmills. This will enable the economics of scale to set in, driving down the cost per unit.
4) Provide mass transit and bike lanes to reduce use of cars.
5) Subsidize research into new or improved clean energy technology, such as new types of solar panels or batteries. The new technology will speed the reduction in cost and improvements in quality.
6) Reduce costs of solar and wind by streamlining procedures for permitting clean energy. In many states of the US, the cost of permits, architectural reviews, etc. is actually a major part of the cost of installing solar or wind. In many cases, it is necessary to get multiple permits from several levels of government to install solar on your roof. Some states have cut the cost of solar installations by thousands of dollars by reducing the paperwork required to install solar on your roof. Such streamlining reduces costs to both the government and people who want to use clean energy. Making the process free for clean energy (no government fees) would also cut costs.
7) Provide educational program to inform people of the benefits to them of installing solar water heaters or electric panels on their homes, cutting energy use by using more efficient appliances, etc.
8) Provide electric car charging stations in many public places. The government can charge a reasonable fee for the electricity, possibly paying the cost of the project, much like a toll road.
9) Providing benefits to people who use electric cars, like reduced registration fees.

10) Establish net metering laws that permit people to sell excess electricity from their home solar panels the electric companies.
11) The government is responsible for providing and maintaining the electric grid. If the government emphasizes running power lines to areas where wind and/or solar are plentiful, it will enable power companies to build wind and solar farms in these locations and distribute the power throughout the country.

These are a few of the ideas for things the government can do to speed the process of converting our fossil fuel energy system a renewable one. There are probably many more things the government can do.

Chapter 27

There is Probably Nothing You Can Do

I have heard the suggestion that the situation is so hopeless, there is no point in even trying. This suggestion sometimes actually comes from the people who would benefit from us not taking any action. Sometimes this idea actually is put forth by the same people who make some of the other claims, contradictory as these two claims may be.

Sometimes it may seem that the situation really is hopeless, and there is nothing you can do. It is true that no matter what we do, global warming and climate change will probably get worse for a while. It will be impossible to stop adding greenhouse gases to the atmosphere overnight. Even if we magically could, there is a certain delayed effect where it takes time for the planet to warm after we add greenhouse gases, much like it takes time for water to boil after you have turned on the heat under the pot. However, global warming and climate change are not yes/no conditions. There are degrees of warming, amounts of sea level rise, levels of how acidic the oceans become, number of acres that will turn into desert, and so on. Every tiny fraction of a degree that we raise the temperature makes it a little worse. That means that every fraction of a degree we reduce global warming is that much less disastrous global warming will be. Every ton of CO_2 or other greenhouse gas we do not release will make life a little more bearable for us and our children. This is why we must not stop trying, even if it seems like a monumental task and we will definitely not be able to stop global warming entirely. Therefore, I would like to end with a few suggestions for things that everyday people can do.

First, you can cut down on your own greenhouse emissions by reducing your energy use. In most cases, this will save you a considerable amount of money in the long run, although it usually involves an initial outlay of cash. You can replace any incandescent light bulbs with LED lights. You can insulate your homes. You can buy high mileage cars when it comes time to replace your car. The same applies to all appliances like air conditioners, refrigerators, etc. Reduce your driving by combining your trips, carpooling, taking public transportation, etc.

You can look into getting solar panels on your roof. In many states, there are companies that will install solar panels at little or no upfront cost and then charge you for the electricity generated by the panels on your own roof. If solar electric is not practical for you, at least look into a solar water heater.

Vote and campaign for political candidates who recognize the danger of global warming and climate change and are willing to fight to reduce it. The League of Conservation Voters is an organization dedicated to scoring politicians on their environmental records. Join this organization to be kept informed on which candidates they have determined are effective at fighting for the environment.

In addition to voting, consider running for political office yourself if you are willing to really get involved. Even at a local level like city government, you can make a difference. A simple thing like installing electric car chargers in public parking lots or providing bike lanes can make a small difference.

There are many charities that fight global warming to which you can contribute. One of my favorites is Trees for the Future (trees.org), an organization that plants fruit trees and other useful trees in underdeveloped countries in order to help feed starving people in those countries. Although their primary purpose is to feed people, planting such trees does greatly help to remove CO_2 from the air and thus reduce global warming. They plant about 10 trees for every dollar they receive in donations. I estimate that it takes about 1,000 trees to absorb the amount of CO_2 that the average American generates every year. That means that donating $100 to this organization once will offset your carbon emissions. This does not mean that you should make one donation to this charity and then forget about global warming, but it certainly is a nice start.

Other charities include Union of Concerned Scientists, The Sierra Club, Environmental Defense Fund, Natural Resource Defense Council, 350.org, Friends of the Earth, EarthJustice, and League of Conservation Voters. This is not an exhaustive list, but it should provide enough organizations for you to contribute to if you want to. Joining these charities with a membership donation not only helps them fight global warming, but it is also an excellent way for you to keep informed. Some of them may help you get involved in local campaigns against pollution if you like.

Recycle. Waste less.

These are just a few ideas. There are other ways you can help. Be aware. The future is in our hands. For more information on how global warming and climate change can be stopped, I recommend the following books:
1) Drawdown: The Most Comprehensive Plan Ever Proposed to Reverse Global Warming
2) Climate of Hope: How Cities, Businesses and Citizens Can Save the Planet
3) Two Percent Solutions for the Planet
4) One Shot: Trees as our Last Chance for Survival
5) You Can Prevent Global Warming (and Save Money!): 51 Easy Ways

If you have any questions about this or any other subjects in this book, I can be contacted at Leithauser@aol.com.

www.ingramcontent.com/pod-product-compliance
Lightning Source LLC
Chambersburg PA
CBHW070159230526
45471CB00002B/727